Energy and Mobility in Smart Cities

ice
Institution of Civil Engineers
Publishing

Energy and Mobility in Smart Cities

Edited by

William J Nuttall
School of Engineering and Innovation,
The Open University, Milton Keynes, UK

David V Gibson
IC² Institute, The University of Texas at Austin, USA

Dariusz Trzmielak
Department of Marketing, University of Łódź, Poland

Alejandro Ibarra-Yunez
Graduate School of Business and Leadership (EGADE),
Tecnológico de Monterrey, Mexico

Published by ICE Publishing, One Great George Street, Westminster, London SW1P 3AA

Full details of ICE Publishing representatives and distributors can be found at: www.icebookshop.com/bookshop_contact.asp

Other titles by ICE Publishing:

Modelling Pedestrian Movement
AECOM (Urban Space). ISBN 978-0-7277-6382-2
Underground Spaces Unveiled: Planning and creating the cities of the future
Han Admiraal and Antonia Cornaro. ISBN 978-0-7277-6145-3
BIM in Principle and in Practice, Third edition
Peter Barnes. ISBN 978-0-7277-6369-3

www.icebookshop.com
A catalogue record for this book is available from the British Library

ISBN 978-0-7277-6425-6

ICE Publishing is a division of Thomas Telford Ltd, a wholly-owned subsidiary of the Institution of Civil Engineers (ICE).

Cover photo: Map pin flat above cityscape and network connection concept

Commissioning Editor: Michael Fenton
Production Editor: Madhubanti Bhattacharyya
Market Development Executive: April-Asta Brodie

Typeset by Academic + Technical, Bristol
Index created by Laurence Errington
Printed and bound in Great Britain by Bell and Bain, Glasgow

Contents

About the editors

Professor William J Nuttall
Professor of Energy at the School of Engineering and Innovation, The Open University, Milton Keynes, UK.

Dr David V Gibson
Senior Research Scientist Emeritus at the IC^2 Institute, The University of Texas at Austin, USA.

Professor Dariusz Trzmielak
Associate Professor, Department of Marketing, University of Łódź, Poland

Professor Alejandro Ibarra-Yunez
Professor of Economics and Public Policy at the Graduate School of Business and Leadership (EGADE), Tecnológico de Monterrey, Mexico.

About the contributors

Dr Peter Boait, PhD, BSc is a Visiting Research Fellow at the Institute of Energy and Sustainable Development at De Montfort University, and also chair of Gloucestershire Community Energy Co-Operative. His research interests are in smart control of domestic energy systems and their system integration for efficient operation at community and national scale.

Professor Richard Bull, PhD, MA, BA (Hons) is Deputy Dean of the School of Architecture, Design and the Built Environment at Nottingham Trent University, UK and Professor of Energy and Behaviour Change. Professor Bull has been a Principal Investigator and Co-investigator on a range of UK and EU research projects exploring innovative participatory methods of engagement surrounding energy use in non-domestic buildings and organisations.

Dr Sally Caird, PhD, MEd, BA(Hons), FHEA has worked in research and teaching roles at The Open University, UK since 1991. She is a Senior Lecturer in Technological Innovation Management and has extensive research experience in examining innovation processes associated with the design, use and impact of low carbon products, services and systems in the built environment.

Professor Matthew Cook, PhD, MSc, FRGS has been employed in urban planning practice and academia for over thirty years. His research interests are in urban innovation and he has been Principal Investigator on several projects concerned with the development of smart cities, urban transport and energy systems. He has also established and now leads the Future Urban Environments research group at the Open University.

Dr Ken Dooley, PhD, BE (Hons) is a Technology Director at Granlund. He has over 15 years' experience as a consultant in the built environment and has worked in London, Sydney, Dublin and Helsinki. He is also a Postdoctoral Researcher in Department of Electrical Engineering and Automation at Aalto University, Finland.

Tahrir Jaber, MSc, is a doctoral candidate in the School of Business and Economics at The Arctic University of Norway – UiT. Her recent education background is in Business Creation and Entrepreneurship. Her current research interests focus on the sustainable business movement and more specifically on how companies/individuals constitute their sustainable concerns and successfully launch sustainable projects.

Dr Remigiusz Kozlowski, PhD Eng, MA, at first worked in technology companies and then in universities. He is engaged in solving the problems of transport logistics in cities using intelligent transportation systems, emerging from over twenty years of work on the subject of the application of modern technologies.

Dr Muhammad Mazhar, PhD, MSc, BE is a Senior Lecturer in Sustainability in Nottingham Business School at Nottingham Trent University, UK. His research interest is in energy/carbon management, low carbon smart cities and citizen engagement.

Elin M. Oftedal, PhD, MSc, BBA is an associate professor in the School of Business and Economics at The Arctic University of Norway – UiT. She received her PhD in Management, with focus on entrepreneurship and innovation. Her research interests include entrepreneurial universities, student entrepreneurship, and sustainable and responsible innovation. She has developed several research applications for different agencies and is currently co-managing an international research project on responsible innovation across six countries funded by Norwegian Research Council.

Dr Stuart Parris, PhD, MSc, BSc (Hons), is Senior Lecturer in the Faculty of Arts and Social Sciences at The Open University. He has researched innovation processes and entrepreneurship at the regional and firm level for over a decade, as well as conducting related consultancy projects for public and private sector organisations.

Professor Stephen Potter, PhD, BSc is Emeritus Professor of Transport at the Open University. His research has explored the human and institutional aspects of transport innovation, the diffusion of cleaner vehicle technologies, low carbon transport systems and more sustainable travel behaviours. Research projects on transport for smart cities have included the Milton Keynes 'Plugged in Places' programme and 'MK:Smart'.

Dr. Richard Snape, MEng MA(Cantab) PhD is a Senior Lecturer in Engineering Science and Mathematics. He worked as an engineer and manager in industry for eight years before returning to academia to research sustainability in the energy sector. His research interests are in modelling smart energy management systems, particularly at community scale.

Sheridan Tatsuno is a Co-Founder of OneReality.com and Principle at Dreamscape Global. He is a Silicon Valley strategist, business author and advisor supporting development across a range of smart technologies, including VR, particularly in support of eco-city initiatives.

Dr Alan-Miguel Valdez, PhD, MA, BSc has dedicated ten years to the study of experimental spaces in the English new town of Milton Keynes that have facilitated the development of new knowledges about sustainable innovation. Through this research he has developed an interest in the ways in which such knowledges coalesce, and travel and in the ways actors make sense of innovative technologies to develop their own narratives about the future of their city.

Acknowledgements

We would like to express our thanks to all those who generously contributed to peer reviewing the chapters of this book. While the editors provided some peer reviews, we are especially grateful to the following: Professor Witold Bartkiewicz, Analysis and Intelligent Systems Department, University of Łódź, Poland; Mr Nicholas Bowker, Independent Engineer, UK; Professor Rene Cabral-Torres, EGADE Tec de Monterrey, Mexico; Professor David Eaton, LBJ School of Public Affairs, University of Texas at Austin, USA; Dr Malgorzata (Margaret) Grzegorczyk of Nottingham Business School, UK and the University of Łódź, Poland; Dr Brendan Hill, Independent Academic Consultant, UK; Dr James Jarrett, Senior Research Scientist, IC^2 Institute, University of Texas at Austin, USA; Dr Bruce Kellison, Director of Bureau of Business Research, IC^2 Institute, University of Texas at Austin, USA; Professor Piotr Niedzielski, Faculty of Management and Economics Services, University of Szczecin, Poland; Professor Anna Pamuła, Chair of Computer Science Department, IT Governance and Service Management, University of Łódź, Poland; and Mr Sheridan Tatsuno, OneReality.com, USA and Sweden.

We are grateful to the International Organising Committee and the Local Organising Committee of the 15th International Conference on Technology Policy and Innovation (ICTPI), hosted by The Open University, UK, in June 2015. In particular, we are most grateful to Kevin Gowans and to Paula Sparling for their assistance with the organisation of the conference, and we acknowledge the kind assistance of the administrative staff of the School of Engineering and Innovation at The Open University in delivering a successful gathering. That conference provided the initial impetus for this book. Without the imagination and support of the ICTPI family this book would not exist.

We are grateful to our commissioning editor at ICE Publishing, Michael Fenton. His thoughtful suggestions and unflinching support are much appreciated. We are also most grateful to Angela Walters at The Open University, who kindly assisted with preparation of the final manuscript.

Finally, we would like to express our most sincere thanks to Dr Wendy Maples, who acted as project manager as the book was assembled. Her conscientiousness and attention to detail were key to ensuring that the book came together successfully.

Introduction

ice Publishing

Nuttall, William J, Gibson, David V, Trzmielak, Dariusz and Ibarra-Yunez, Alejandro
ISBN 978-0-7277-6425-6
https://doi.org/10.1680/emsc.64256.003

Chapter 1
The smart future

William Nuttall
The Open University, Milton Keynes, UK

Alejandro Ibarra-Yunez
Tecnológico de Monterrey, Mexico

Dariusz Trzmielak
University of Łódź, Poland

David Gibson
The University of Texas at Austin, USA

1.1. Introduction: the smart city transition in context

In this book we bring together a range of expert perspectives on a topic of global importance – the changing nature of communities in the digital age and the implications that follow for mobility and energy supply and demand, mainly in urban settings. This is indeed an important change, but, of course, society has always been in a state of change and cities have long been a key driver of that process. Now, however, there is a pervasive sense that something more substantial and profound is underway; perhaps among the most profound changes in human history. We are not suggesting that the shift will come in the next 5 years, we are saying that it will take closer to 100 years and, importantly, that we are already well into the journey. In this book we consider a somewhat shorter time horizon in recognition of the deep uncertainties and possibilities for divergent pathways that currently exist.

Seeking perspective on the profound changes ahead, it is perhaps helpful to take a bold step back and reflect on the issues at the largest possible scale. In that spirit we shall briefly see the issues in terms of the development of the human experience. In prehistory, human beings emerged as the dominant species on the planet as a consequence of some fundamental shifts. These included the ability to make and control fire and the shift from hunter-gatherer living to agriculture, including the associated focus on settled living. The invention of writing, and later of the printing press, facilitated the movement of knowledge from one town to another, but it also greatly boosted the sharing of information between generations. The later invention of the telegraph and then of the telephone permitted more rapid communication between otherwise distant communities. The innovation of electronic communication gave a further boost with the development of

Figure 1.1 Graphical representation of Ecocity Standards. (Adapted from: Ecocity Builders and the International Ecocity Advisory Committee, 2019)

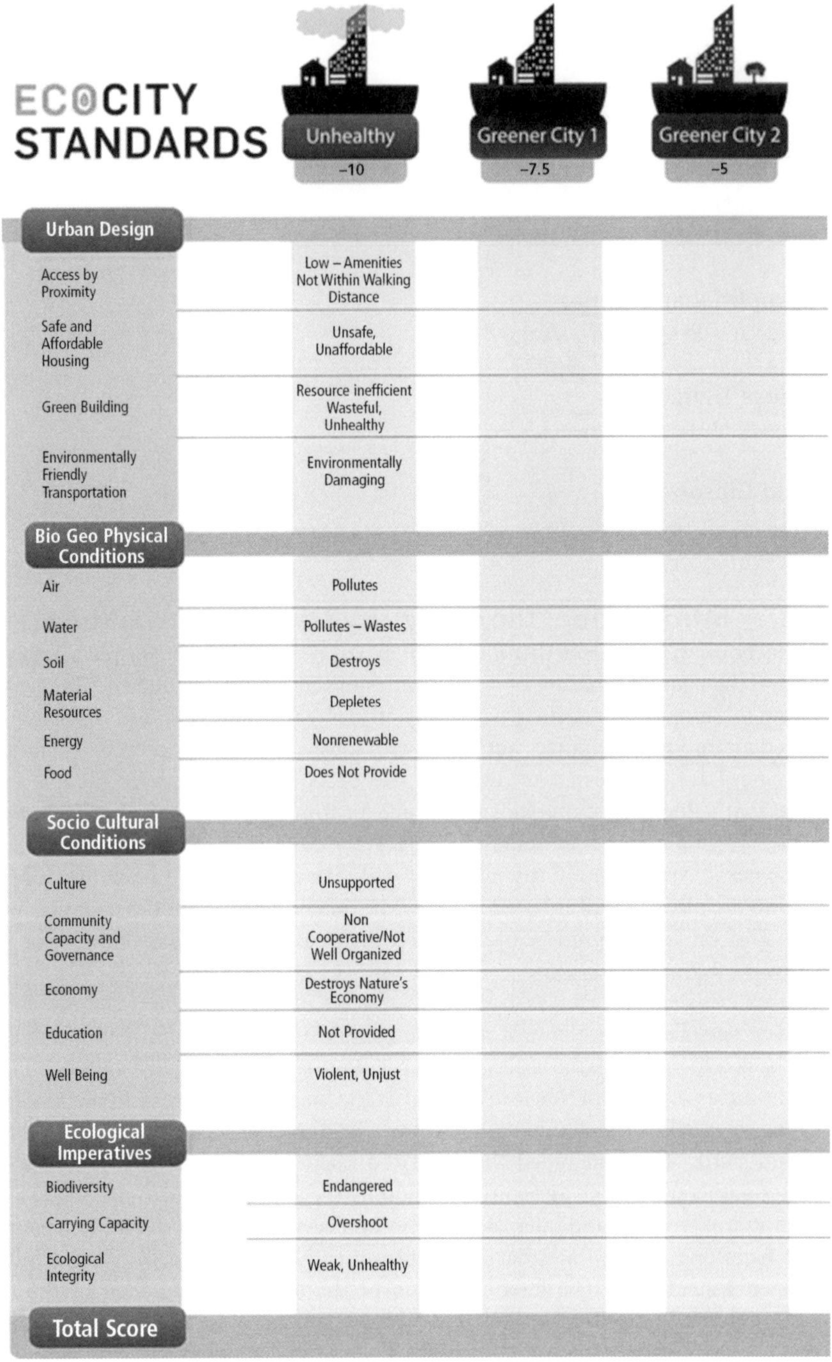

Greener City 3 −2.5	Eco City 1 Ecocity Standard 2.5	Eco City 2 5	Eco City 3 7.5	Gaia Level 10
	Walkable, Accessible			Complete + Sustainable
	Safe, Affordable			Safe, Affordable
	Resource efficiency, Healthy			Regenerative
	Does not Damage			Improves Environment
	Clean			Purifies
	Clean and Safe			Purifies
	Healthy			Restores
	Responsible			Sustains
	Clean and Renewable			Clean and Renewable
	Healthy and Accessible			Nutritious and Abundant
	Healthy, Supported			Nurtured
	Health, Participatory			Highly Organized/ Highly Cooperative
	Healthy and Equitable			Restores Nature's Economy
	Lifelong, Accessible			Provide for All
	Quality of Life Satisfaction			Justice, Peace & Contentment
	Healthy			Sustains
	Low Impact			Within the Biosphere's Limits
	Healthy			Strong, Restorative

packet switching (1967) and internet protocols (in the 1970s). The revolutions enabling communication over greater and greater distances have, however coexisted with a trend for agglomeration in cities, rather than a dynamic towards dispersion. In addition, concern for climate change, which is mainly caused by carbon dioxide emissions, together with a wider range of environmental degradations, is prompting a range of innovations associated with a low-carbon future, including energy storage (e.g. see Chapter 8) and new energy vehicles innovations (e.g. see Chapter 2). Such eco-innovations occupy a major part of industry development at the beginning of the 21st century.

The first wave of architectural and urban planning ideas arising in response to the environmental stresses of late 20th century life came together in the 1980s under the label 'eco-cities', a term reportedly first coined by Californian architect Richard Register in 1987 (Rapoport and Vernay, 2014). Register established Ecocity Builders, a non-profit organisation dedicated to ecological principles for urban living. Ecocity Builders has helped to develop a framework and standards initiative by which the concept of the eco-city can be promoted, and the performance of eco-cities can be measured (Figure 1.1).

The concept of the eco-city emerged from environmentally conscious thinking in and around Berkeley, California through the 1970s, partly as a result of the actions of a group of early pioneers, including Register, who had founded a group, Urban Ecology. The concept of the eco-city gained international attention, and arguably instances can be seen in places as diverse as

- Dongtan Eco-City, Shanghai, China
- Masdar City, Abu Dhabi, United Arab Emirates
- Sonoma Mountain Village, California, USA
- Hammarby Sjöstad, Stockholm, Sweden
- EcoVillage at Ithaca, New York, USA
- Malmö Western Harbour, Sweden.

These communities have been selected for investigation by Rapoport and Vernay (2014), who concur with other commentators when they observe that 'as the usage of the term spreads, what exactly constitutes an eco-city seems to be even more unclear'. These authors use discourse analysis to gain insight into such issues. We note the importance of Sweden to the emergence of eco-city thinking, and these issues are considered explicitly in Chapter 4.

In the early years of the 21st century, interest in, and concern for, environmental sustainability increased in most parts of the world. There was, however, another shift also underway in this same period – the emergence of digital technology, especially as applied to smartphone technology. The global rollout of smartphones has been quite remarkable. In 2018 the market research firm Newzoo predicted that the 2020s would see more than 3.5 billion smartphones worldwide (Figure 1.2) (Takahashi, 2018).

In the first decade of the 21st century the term 'smart city' emerged to help describe the emergence of cities defined by the new digital realities. It soon became clear that the

Figure 1.2 Newzoo's estimates for global smartphone users, by region. (Adapted from Takahashi, 2018)

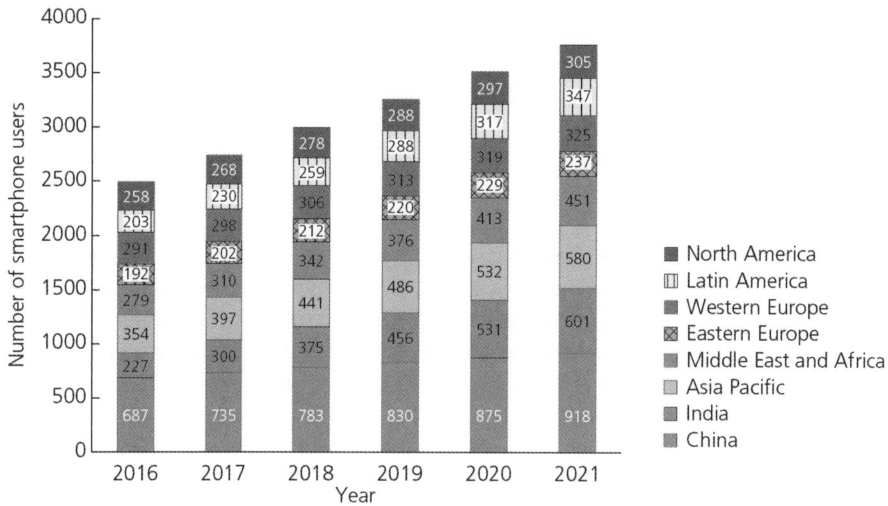

relationship of new digital communication tools with innovations in energy supply and urban mobility could facilitate progress in key areas of concern to eco-city advocates. For many people the smart city vision became a means by which eco-city goals might be met. This book sees the smart city in such terms, but we take the view that the concepts of the eco-city and the smart city remain distinct. The focus here is on the smart city, but many of the challenges to be faced are environmental and ecological.

The smart city has the potential to combine new technologies with human potential and more collaborative governance. It can leverage high-bandwidth internet communications (both wired and wireless) with potentially distant cloud-based data storage and software services to provide joined-up services across key domains of urban life, including in the areas of mobility (public, private and of goods) and energy (in all its forms). Example technologies include smart electricity distribution grids and self-driving cars. This joining-up of services provides for improved system efficiency, reliability and robustness. The use of modern information and communications technology (ICT) and the further development of system artificial intelligence (AI) provide the prospect of decision-making across a breadth of data and with a complexity of considerations that no individual human being, or even a human-mediated system, could ever hope to match. Stanford University researchers have forecasted significant impact of AI on eight industries (AI100, 2015): it is noteworthy that they find that the most immediate impact of AI is likely to be in the areas of transport and mobility.

Ogie *et al.* (2017) have listed the following areas where benefits can be expected from the development of smart infrastructure

- self-monitoring and accuracy in decision-making

- efficiency and cost savings
- reliability
- security, safety and resilience
- user interaction and empowerment
- sustainability
- redundancy minimisation
- response time
- low-carbon dioxide footprint ('carbon footprint')
- service quality.

1.2. Security of the smart city

The smart city is usually seen as a positive step forward. The policy and economic discourse focuses on enhanced efficiencies and improved quality of life for citizens. As noted earlier, the smart city has the potential to assist in addressing the global challenges associated with greenhouse gas emissions and anthropogenic climate change. In such respects the smart city has the potential to mitigate risks. The question must be asked, however: does the smart city also increase security risks for citizens and the economy? There is relatively little written officially on smart city and security, but the UK has taken the lead in one respect by developing a formal Publicly Available Specification (PAS). The UK government's Centre for the Protection of National Infrastructure (CPNI) has developed PAS 185:2017 in collaboration with the British Standards Institution and the UK government's Department for Business, Energy and Industrial Strategy. Entitled 'Smart cities – Specification for establishing and implementing a security-minded approach' (BSI, 2017), the specification emphasises the importance of security awareness and preparedness as smart city technologies roll out and start to transform the urban experience. It is important to note that innovations that can do so much to empower benign actors in civil society can open up the risk of giving new capabilities to those intent on causing harm and disruption. By simply remembering and pausing to consider such risks, the chance of being caught unprepared is reduced.

1.3. Utopias, dystopias and radical thinking

The digital revolution is enriching the human experience and providing new capabilities that would barely have been imagined 50 years ago. Norman Jewison's 1975 dystopian science fiction film *Rollerball* depicts the year 2018, with the framework that it is a world of multi-channel ultra-violent TV with no mobile phones at all. But the film does depict a powerful computer, Zero, which acts as the repository of all human knowledge. Arguably, such a repository exists today, but unlike in the movie we can all access it directly from the smartphone in our pocket. Few in 1975 would ever have imagined such a possibility.

In his thought-provoking book *A Rough Ride to the Future,* James Lovelock lays out the case that planet Earth faces a difficult existence given one key reality (Lovelock, 2014): thehe dominant species on the planet is a tribal, carnivorous primate. He suggests that, despite relatively high individual intelligence, human beings will not be capable of ensuring planetary sustainability. It simply runs counter to the nature of the species. Notwithstanding the rather bleak pessimism inherent in such a view, Lovelock does hold out one small glimmer of hope. Thanks to the human beings' silicon-based innovations in ICT and AI they might achieve a shift in decision-making away from self-interested individuality towards a more collective and sustainable path. Essentially, this would appear as a shift away from the individualised intelligence characteristic of mammals, and primates in particular, to a hive intelligence for all humankind. The human experience might start to resemble that of the social insects. We will have city-hives made possible by advanced ICT. Technology is changing society. This has happened at many times in the past, but the changes are particularly profound and fast-paced at present.

Lovelock goes further to posit that humanity should end the trend of millennia whereby humanity seeks to understand and control nature. Humanity should aim to detach itself as far as possible from the natural world. As concerns the air we breathe, we should merely seek to control the atmospheric environment within the cities of the future and 'we should give up on trying to save the planet' (Sample and Smith, 2014). As with another dystopian science fiction film of the 1970s, *Logan's Run*, humanity should live in sealed city bubbles. Only by such separation from the planetary ecology will humanity be able to save it. His point centres on the notion that human beings delude themselves if they believe their existing approaches to planetary-scale actions will ever be helpful. We should retreat from such ambitions, behave more like the social insects and leave the planet to its own devices. If we can leave the Earth alone, it will be fine. Nature will take care of the countryside and wild spaces if only the human species can resist the temptation, however well intentioned, to intervene. Such ideas relate closely to an environmental philosophy known as ecomodernism (Asafu-Adjaye *et al.*, 2015). Of course, the embryonic smart cities of the first quarter of the 21st century are still a long way from being Lovelock's hive-intelligent systems. Any journey to such a future will be contested and circuitous. Already the emerging realities are colliding with long-established norms and expectations.

Noting Lovelock's provocative observations, we must at this point concede that this book does not go to such philosophical extremes. We are not suggesting that the smart cities of the future will be isolated communities under glass, but we do point to the idea that high-density living separated from planetary ecology can be consistent with ecological responsibility. In so doing we present a vision very different from the eco-cities approach mentioned earlier. Our focus in this book is grounded in current realities and looks to near- to medium-term innovations within broadly similar social contexts. As such, we take a roughly 30-year view and look to the year 2050.

The term 'smart city' covers many aspects of urban innovation and is, partly consequently, open to differing interpretations and meanings. In 2012, Batty *et al.* provided a most helpful analysis of the fundamental functions of a smart city: this is summarised in Table 1.1.

The prominence given by Batty *et al.* to people, governance and the quality of life is important. This book combines consideration of the technically grounded issues of mobility and energy services with the more social and political issues of governance and communities. In this book we do not present the smart city as a city machine to be

Table 1.1 The six core functions of a smart city (Batty *et al.*, 2012)

SMART ECONOMY (competitiveness)	**SMART PEOPLE (social and human capital)**	**SMART GOVERNANCE (participation)**
■ Innovative spirit ■ Entrepreneurship ■ Economic image and trademarks ■ Productivity ■ Flexibility of labour market ■ International embeddedness ■ Ability to transform	■ Level of qualification ■ Affinity to lifelong learning ■ Social and ethnic plurality ■ Flexibility ■ Creativity ■ Cosmopolitanism/open-mindedness ■ Participation in public life	■ Participation in decision-making ■ Public and social services ■ Transparent governance ■ Political strategies and perspectives
SMART MOBILITY (transport and ICT)	**SMART ENVIRONMENT (natural resources)**	**SMART LIVING (quality of life)**
■ Local accessibility ■ (Inter-)national accessibility ■ Availability of ICT infrastructure ■ Sustainable, innovative and safe transport systems	■ Attractivity of natural conditions ■ Pollution ■ Environmental protection ■ Sustainable resource management	■ Cultural facilities ■ Health conditions ■ Individual safety ■ Housing quality ■ Education facilities ■ Touristic attractivity ■ Social cohesion

optimised. Rather, we take the view that the smart city of the future will be a socio-technical system, and the journey to that future will be a politically and economically contested one.

1.4. Looking for smart cities

The Cities in Motion project led by the University of Navarra in Spain provides a very interesting overview of the multi-dimensional nature of change in the modern city. Adopting a similar approach to Batty *et al.* (2012), the Spanish-led project considers the following: governance; urban planning; technology; the environment; institutional outreach; social cohesion; mobility and transportation; capital; and the economy (IESE Business School, 2019).

Around the world we see a set of cities that have emerged from relative obscurity over the last 50 years to become globalised centres of commerce, trade, aviation and the knowledge economy. Prominent examples include Dubai and Singapore. In many ways they exemplify the smart city, as they seek to deploy cutting-edge solutions in infrastructure and public services, but that does not mean that they are smart cities in the sense explored in this book.

Such cities as Singapore and Dubai are certainly 'high bandwidth' – an attribute shared by many other world regions. This attribute would appear to be necessary for a city to be smart, but equally it would appear to be insufficient. A smart city must in some sense be a modern city and it should arguably have a global orientation – serving the world while simultaneously meeting the needs of its inhabitants. While we note that interest in smart cities is global – see, for example, the moves to smart urban traffic management in Chennai, India, announced in 2018 (*Times of India*, 2018), the reality is that most instances of cutting-edge innovation in this sector are to be found in developed countries. It is arguable that, in fact, the best examples of smart city innovation are to be found in cities that have seen significant economic growth in recent decades. With such thoughts in mind, Singapore and Dubai exhibit many of the attributes of a smart city and combine them with a very high proportion of expatriate workers. The cities are fixed in geography, but they are growing fast, and in so doing they bring in global talent. They are truly global cities. In 2016, Singapore's population comprised 61% citizens with 9% permanent residents and 30% non-residents (Yang *et al.*, 2017), while for the United Arab Emirates as a whole it is estimated that citizens form only 12% of the population (CIA, 2017). It is interesting to note that governance and law enforcement are somewhat more authoritarian in these territories than is the norm for the established cities of Europe and North America. Is such control and oversight inevitable in the smart city? Is it a key component of the social harmony and prosperity seen in these places? A smart city clearly enables an efficient and productive life for its citizens – but when combined with a desire by the state to have deep insight into the behaviours of citizens, a darker side of the smart city reality may become apparent. The fact that chewing gum is effectively illegal in Singapore, or that drinking on the subway is a punishable misdemeanour, is arguably less interesting than the ability of the state to prove its case in such examples, via closed-circuit digital video capture. In Dubai, similar capabilities can be deployed against public drunkenness and indecency. To be sure, a tidy and efficient life can be a pleasant life – but it leads one to ask what the impacts are upon creativity and innovation. These cities are high tech, but also somewhat authoritarian and hierarchical in respect of political governance and national culture. Arguably, the true smart city is more creative and less managed. As regards the validity of such an opinion, only time will tell.

Will citizens of the future smart city be aware of the surveillance and information-gathering that is taking place? Will citizens be able to understand and use the capabilities of new systems and infrastructures? Such questions have motivated work into 'digital transparency' such as that described undertaken by Side Walk Labs of Toronto, Canada, with its Designing for Digital Transparency in the Public Realm project (Side Walk Labs, 2019). This aims to use digital technologies to make city infrastructures and amenities easier to perceive, understand and use.

On a first impression, the emergent cities of Singapore and Dubai feel closer to the future than the historic cities of Europe, and it must be conceded that security surveillance is very much part of the lived experience in London in the 21st century. Singapore and Dubai aim to be among the first to adopt innovative new concepts – for example, both Emirates and Singapore Airlines were early adopters of a new approach to civil aviation promoted by Airbus in the form of the Airbus A380 super-jumbo aircraft. In that case, however, large-scale adoption did not occur, and closure of the A380 project was announced in 2019 (Blum, 2019). It is interesting to compare that aim with the goal to develop these innovations. It seems that a smart city does not need to be a research city, notwithstanding the fact that Singapore is home to some globally competitive research universities. Going further – whether the smartness of a city might even be an inhibitor of creativity and innovation is an outcome that remains to be seen.

In addition to the globally important hubs of Singapore and Dubai, there are also nationally and regionally important smart cities catering to a more local commercial and industrial base. Examples of such cities would include Austin, Texas, in the USA and Milton Keynes in the UK. Milton Keynes features strongly in this book, especially in Chapters 2 and 5. Many of the contributions to this book arose originally from a conference held in Milton Keynes in the summer of 2015. It was the 15th conference in a series: the International Conferences on Technology Policy and Innovation.

Just as with Singapore and Dubai, cities such as Austin and Milton Keynes have seen dramatic change and growth over the last 50 years, driven by emerging innovations in information technology (IT) and logistics, respectively. These cities are emerging as smart cities, but they have also played a major role in regional and national innovation systems. In the case of Austin, the city has been at the heart of a knowledge cluster built around a world-class and growing research university – the University of Texas at Austin – supported by numerous other universities and colleges in the region. As a 'college town', Austin is indeed a centre of radical creativity and innovation, famous not only for developing new IT solutions and business models but also for a particularly lively musical and cultural scene. Milton Keynes' claim to be a smart city is based on a somewhat different proposition – as a place where change is relatively easy and where the instinct of decision-makers is to say 'yes'. It plays such a role near equidistant between the capital, London, and two major knowledge centres, Oxford and Cambridge, both famous for their research universities and the constraints of planning and space that come from a medieval heritage. Recently, the road transport links between Milton Keynes and these knowledge centres have improved significantly, and improvements in east–west transport infrastructure through Milton Keynes remain a national strategic priority for the UK (NIC, 2017). As a consequence, Milton Keynes will continue to be an interesting case study of the UK smart city experience.

1.5. The structure of the book

This book is presented in four parts. The first part is devoted to an exploration of *communities*, and it draws heavily on the European experience. We give special mention to communities to give emphasis to the point that cities are about people more than they are about technologies. Cities are where people increasingly are choosing to live. The city is a home and it is a community. Part 1 reviews and examines leading European developments as established urban communities confront the digital age in times of ecological concern. As we look to the future, we note the maxim, most frequently attributed to William Gibson, that goes, 'The future has arrived, it's just not evenly distributed yet.'

Chapter 2 by William Nuttall considers similarities and differences seen between two historic European capital cities (London and Paris) as they seek to synergise policy around mobility and energy in the new digital environment. The chapter also looks at some wider British initiatives examining the interface between the new increasingly electric mobility technologies and the established electricity distribution system.

Chapter 3 by Stuart Parris examines the relationship between the emergence of smart cities and the prospects for environmental entrepreneurship. The extent to which city smartness, innovation and environmental action are correlated is an important issue to unpick if one is to appreciate properly the potential impacts of smart city emergence.

Chapter 4 by Sheridan Tatsuno examines further the ways in which smart cities can be expected to help the world meet the need to greatly reduce greenhouse gas emissions. He particularly focuses on the experience of various Swedish cities in confronting the challenges and obligations of global policy. In many ways, Sweden represents a beacon that other countries might be guided by.

The second part of the book deals with *mobility*, and a future that is increasingly electric and autonomous, building on Chapter 2.

Chapter 5 by Stephen Potter, Alan-Miguel Valdez and Matthew Cook introduces the emergence of new mobility options and examines them holistically as disruptive socio-technical systems. This is a sector full of novelty and innovation, but yet in many ways the fundamentals are unchanged. The chapter presents a historical perspective and explains how innovation is welcomed and resisted by various stakeholders. In this book, as we look to the future, we do not imagine that the future is a corridor that lies ahead. Rather, we see a range of uncertainties and opportunities.

Chapter 6 by Dariusz Trzmielak and Remigiusz Kozlowski presents an overview of technological innovation in the sectors of concern before providing an illustrative case study from central Europe – the City of Łódź. Historically, the city was a major centre of the international textile industry, but in recent decades it has been in transition, with a diversifying industrial and commercial base. As the chapter describes, the city has enthusiastically identified transport infrastructure improvement as a key priority, with major civil engineering works having been undertaken in the city centre.

The third part of the book focuses on *energy*. In particular, we look beyond smart metering and smart grids to consider electricity markets in transition economies, and we examine the role of renewable energy in future smart energy systems.

Chapter 7 by Alejandro Ibarra-Yunez examines the Mexican experience with electricity policy and economics. Electricity market design is considered with a view to its implications for climate change mitigation policy and a changing electricity system.

Chapter 8 by Richard Snape and Peter Boait shows the power of agent-based modelling as a technique by which energy futures can be assessed. In particular, this chapter considers the potential role of renewable energy in future communities – reminding us of the importance of the demand side of energy issues and the role of individual citizens in our energy future.

The fourth part of the book builds upon the ideas in Chapter 8, to consider explicitly matters of *governance*. Here we delve into the political issues shaping the smart future, and we note the demands of sustainability, rights and justice. We also note the role and power of different stakeholders in times of change, be they energy consumers, city governments or large, established energy companies. We see a way ahead in which these stakeholders will evolve in terms of power and interests, especially in the face of growing environmental concerns and the power of digital systems and artificial intelligence.

Chapter 9 by Sally Caird sets out issues relating to the evaluation of cities in respect to existing and emerging policy goals. How is change best achieved and how might governance itself be smarter?

Chapter 10 by Jaber Tahrir and Elin Oftedal considers the role of corporations in our energy future. The chapter presents a case study of the Norwegian-headquartered energy company Equinor. The study considers the economic, environmental and social challenges facing the sector and how Equinor manages and seeks to lead sustainability change.

Chapter 11 by Richard Bull, Ken Dooley and Muhammad Mazhar considers the engagement of the individual with the smart city of the future. In particular, a distinction is drawn between the individual as an energy consumer and as a political actor. Individuals see energy issues in these different ways at different times. Can future governance better handle these realities and use the new smartness to improve policy-making and system operation?

The book is concluded by Chapter 12, prepared by the editors.

The core content of this book (Chapters 2–11) has been peer reviewed and subject to editorial oversight. The ideas expressed represent those of the original chapter authors and do not necessarily represent the opinion of the editors. This book is published by ICE Publishing, but should not be regarded as a statement of any institutional position or opinion.

There are many ways to define the smart city, and in this book we do not seek to present or establish a single prescriptive definition. We instead seek, via expert perspectives and a diverse range of research methodologies, including market design economics and agent-based simulation, to explore the issues as they appear at the end of the second decade of the 21st century. One thing, however, we do seek to stress, and despite mentioning it earlier, it merits repetition: cities are about people and they are not narrow technical systems. They are socio-technical systems in the fullest sense. For that reason, we give significant emphasis to matters of community and governance, as well as to mobilities and energy.

We see a landscape of choice and flexibility in which technology can help provide for enhanced safety, security, sustainability and prosperity; however, we also see a future landscape where disruption and change must be expected, but in ways that cannot be predicted. The future comprises a set of possible scenarios; it is not determined in any sense, and most definitely it is not technologically determined.

We hope that readers will find the chapters that follow both interesting and informative. We would welcome receiving comments from readers by whatever means is most comfortable. The editors' contact details are readily available online.

Acknowledgements

We are most grateful to Professor Rohinton Emmanuel of Glasgow Caledonian University, UK, for sharing his insights and experience. Any errors and omissions are, of course, the responsibility of the authors alone.

We would also like to thank our commissioning editor, Michael Fenton, for most thoughtful and insightful comments that have greatly assisted us in the preparation of this chapter specifically and the book more generally.

REFERENCES

AI100 (One Hundred Year Study on Artificial Intelligence) (2015) *Artificial Intelligence and Life in 2030*. Stanford University, Stanford, CT, USA. https://ai100.stanford.edu/sites/g/files/sbiybj9861/f/ai100report10032016fnl_singles.pdf (accessed 19/06/2019).

Asafu-Adjaye J, Blomqvist L, Brand S *et al.* (2015) An ecomodernist manifesto. http://www.ecomodernism.org (accessed 19/06/2019).

Batty M, Axhausen K, Fosca G *et al.* (2012) Smart cities of the future. *The European Physical Journal Special Topics* **214**: 481–518.

Blum S (2019) Airbus to stop production of the colossal A380 as airlines transition to smaller planes. *Popular Mechanics*, 14 February. https://www.popularmechanics.com/flight/airlines/a26341068/airbus-a380-production-cancelled (accessed 19/06/2019).

BSI (British Standards Institution) (2017) PAS 185:2017. Smart cities – Specification for establishing and implementing a security-minded approach. BSI, London, UK.

CIA (Central Intelligence Agency) (2017) The world factbook. https://www.cia.gov/Library/publications/the-world-factbook/geos/ae.html (accessed 19/06/2019).

Ecocity Builders and the International Ecocity Advisory Committee (2019) Assessing and guiding progress towards ecologically healthy cities. https://ecocitystandards.org (accessed 24 April 2019).

IESE Business School (2019) Cities in motion. Smart urban governance: rising to challenge. https://www.iese.edu/en/faculty-research/research-centers/cgs/cities-motion-strategies/?_ga=2.149322229.297456314.1540553034–2028846514.1540553034 (accessed 19/06/2019). http://citiesinmotion.iese.edu/indicecim/?lang=en (accessed 19/06/2019).

Lovelock J (2014) *Rough Ride to the Future*. Allan-Lane, London, UK.

NIC (National Infrastructure Commission) (2017) *Partnering for Prosperity: A New Deal for the Cambridge–Milton Keynes–Oxford Arc*. https://www.nic.org.uk/wp-content/uploads/Partnering-for-Prosperty.pdf (accessed 19/06/2019).

Ogie RI, Perez P and Dignum V (2017) Smart infrastructure: an emerging frontier for multidisciplinary research. *Proceedings of the Institution of Civil Engineers – Smart Infrastructure and Construction* **170(1)**: 8–16.

Rapoport E and Vernay AL (2014) Defining the eco-city: a discursive approach. *Reflections* **3**. https://www.westminster.ac.uk/file/42096/download (accessed 19/06/2019).

Sample I and Smith E (2014) James Lovelock: we should give up on saving the planet – video. *The Guardian*, 10 April. https://www.theguardian.com/environment/video/2014/apr/10/james-lovelock-climate-change-saving-planet-video (accessed 19/06/2019).

Side Walk Labs (2019) Designing for digital transparency in the public realm. https://www.sidewalklabs.com/dtpr (accessed 19/06/2019).

Takahashi D (2018) Newzoo: Smartphone users will top 3 billion in 2018, hit 3.8 billion by 2021. https://venturebeat.com/2018/09/11/newzoo-smartphone-users-will-top-3-billion-in-2018-hit-3-8-billion-by-2021/ (accessed 19/06/2019).

Times of India (2018) Traffic management to be part of smart city. *Times of India, Chennai Edition*, 6 December. https://timesofindia.indiatimes.com/city/chennai/traffic-management-to-be-part-of-smart-city/articleshow/66963915.cms (accessed 19/06/2019).

Yang H, Yang P and Zhan S (2017) Immigration, population, and foreign workforce in Singapore: an overview of trends, policies, and issues. *HSSE Online* **6(1)**. https://www.hsseonline.edu.sg/journal/volume-6-issue-1–2017/immigration-population-and-foreign-workforce-singapore-overview-trends (accessed 19/06/2019).

Part 1. Communities

ICE Publishing

Nuttall, William J, Gibson, David V, Trzmielak, Dariusz and Ibarra-Yunez, Alejandro
ISBN 978-0-7277-6425-6
https://doi.org/10.1680/emsc.64256.019

Chapter 2
Historic European cities and the smart future

William Nuttall
The Open University, Milton Keynes, UK

2.1. Introduction

In Chapter 1 we have been introduced to the issues facing the modern city. There is, however, no one single reality to consider, rather a multiplicity of experiences and infrastructures. Around the world one finds an enormous diversity of cities facing local and specific challenges. One can think, for instance, of cities as diverse as Paris, Houston and Mumbai. Such cities differ in terms of urban density, relative affluence one to another, the range and extent of wealth within the city, levels of infrastructure, and the expected balance between public and private investment. Within Europe there is also diversity, but there are also repeated and commonly encountered issues. Many European cities combine affluence, density and the constraints of carefully protected heritage. While European cities often have much in common, within the EU different European countries also have much policy autonomy in areas relating to energy and urban mobility. For example, for member states of the EU energy policy is a shared competence, with responsibilities and powers reserved at both the European and the national level. A well-known example is that while there is an EU-wide set of energy emissions policy goals, the actual mix of electricity-generating technologies remains a firmly national responsibility. In a spirit of subsidiarity, local autonomy exists right down to the municipal level for some aspects of energy policy, especially regarding issues in urban mobility, air quality and traffic congestion. The nexus of energy, mobility and information technology (IT) raises a set of policy issues at all levels but most notably at the most local level – especially for the smart cities of the future.

In this chapter we seek to explore some emerging issues holistically, particularly as they relate to north-western Europe. We seek to assess synergies between the domains of energy, mobility and IT and to understand better how such synergies might shape the trajectory of European urban communities. We shall also introduce the importance of governance, including consideration of the role of regulation and the legitimacy and authority of decision-makers. It is not the intention of this chapter to provide a full overview of the range of technology options relating to the new smart cities, rather it is to introduce the nexus of energy and mobility through a limited set of specific examples. The subsequent chapters of this book will expand on such ideas in greater depth, with other perspectives and with reference to other contexts.

2.2. Paris and London: common challenges and diverse responses

Both Paris and London are historic, affluent, high-density European cities. Both share concerns for efficiency, air quality, safety, social justice and reduced greenhouse gas emissions. While the challenges are similar, even a casual visitor can readily see differences of policy and strategy.

One area of visible difference relates to constraints over the use of private cars in the city centre and how those constraints are imposed. In February 2003 London introduced a technology-mediated road-pricing scheme known as the Congestion Charge. At the time of its launch it was the largest such scheme in any capital city (BBC, 2003). From the outset the scheme was expected to motivate a shift away from private cars towards public transport, such as the London Underground rail system, but more importantly to bus services. Indeed, 300 new buses were purchased as an act in support of such a shift (Beckett, 2003). While the policy was ostensibly concerned with traffic congestion, it soon became apparent that air quality, greenhouse gas emissions and sustainability generally were associated policy concerns of growing importance. As a consequence, fee exemptions were provided for pre-registered cars emitting less than 75 g/km of carbon dioxide and compliant with the Euro 5 emissions standard as established by regulation number 715/2007 of the European Parliament and of the Council (with entry into force on 2 July 2007). The exemption scheme was known as the Ultra-Low Emissions Discount (TfL, 2015).

Figure 2.1 shows the warnings faced by drivers entering the Congestion Charge Zone in London. Specifically, the white letter ‘c’ in a red circle denotes entry into the zone. In this

Figure 2.1 Entry to the London congestion charging zone. (Source: Wikimedia user Beao, licensed under CC BY-SA 3.0)

way, drivers are warned that they must pay. As originally launched, drivers could pay the charge in the hours after their entry into the city by making payment at a retail outlet, such as a shop or a petrol station. In July 2013, however, it was announced that this aspect of the policy would end, and along with it the ability to pay with cash (TfL, 2013). The payment system would become entirely electronic and IT mediated. Consequently, policy had by this point brought together issues in energy (vehicle power sources), mobility (some mobility options exempt or favoured) and digitisation (zone enforcement and payment systems). As such, it was an early precursor of the smart city. These impacts go far beyond the original congestion concerns that motivated the initial policy.

In the years after the introduction of the Congestion Charge, other shifts concerning mobility and energy were seen. For example, London saw a substantial shift in policy and infrastructure in favour of cycling in the city (Figure 2.2). These developments have received uninterrupted and bipartisan political support, as, for example, when political control shifted from the Labour Party to the Conservatives in May 2008 and back again in May 2016.

Meanwhile, on the other side of the English Channel, in Paris, there is no similar congestion charge, and policy restrictions on vehicles have been focused explicitly on air quality concerns. Vehicles are required to register for a Crit'Air badge based on EU emissions standards (typically at Euro 2 or Euro 3 standard) (EU, 2019). The policy is not mediated by technology, such as digital camera data acquisition at the boundary and cross-checked against payment records (see Figure 2.1), but rather is processed

Figure 2.2 Growth of cycling in London 2000–2017. (Source: TfL, 2018)

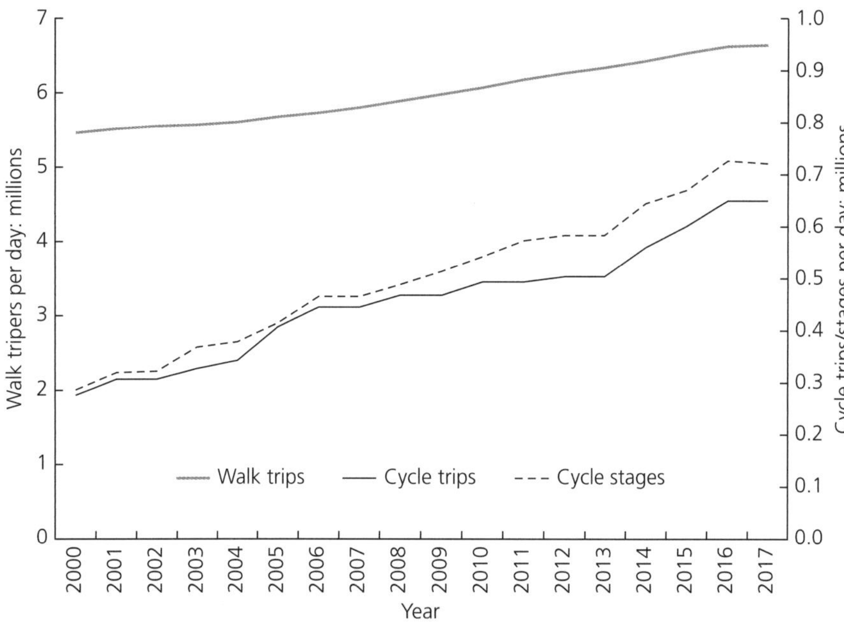

Figure 2.3 (a) Lime electric scooters at a bus stop in Paris, February 2019. (b) Lime electric bicycles on the campus of The Open University, in Milton Keynes, UK, March 2019

(a) (b)

manually by police officers. As such, a visitor does not encounter a visible infrastructure of enforcement, as is seen in London.

A visit to Paris very quickly reveals other differences in the journey towards smart mobility. Paris is the first city in Europe to welcome electric stand-on scooters. In the summer of 2018 the American company Lime led the way with the roll out of smartphone-app rented electric scooters (Lee, 2018) (Figure 2.3). Less than 1 year later, one could see scooters from several competing companies scattered throughout the city. The pace of uptake has been remarkable, while in London such scooters are at the time of writing, still a rare sight indeed. The reason is that anyone riding such a scooter in the UK is risking prosecution and a fine. UK regulations ban hoverboards, electric scooters and similar vehicles from both the roads and pavements (Cook, 2018). These regulations, however, were drafted long before such electric technologies were even imagined. Today, regulatory caution and conservatism stand in the way of a greening and smartening of UK cities.

In this chapter as we examine the future of personal mobility we focus on two concurrent, overlapping, but still separate, trends. One is the electrification of urban mobility for a low-carbon future and the second is the roll-out of digital innovation altering the ways in which the individual achieves mobility. The example of the Parisian electric scooter combines both of these elements – electric motive force with smartphone app-based payments. This is merely one example from a far-wider trend. Let us take a step back to gain some perspective.

The 20th century saw the emergence, starting in the USA, of a dominant paradigm in personal mobility – the individually owned family car. Powered by fossil fuels, the individual's relationship with the technology was to be an owner and a driver. The device was mechanically powerful, but it was polluting and it, for the first few decades at least, had no computational capacity whatsoever. The deployment of digital technologies in cars has had a long heritage, starting with the first electronic engine control units (ECUs) pioneered by General Motors in the USA in the 1980s. Typically, an ECU, as now found in all modern cars, provides for efficient fuel injection and monitors vehicle systems for faults. The power and sophistication of on-vehicle electronics have developed rapidly over recent decades, and these systems increasingly started to involve aspects of driving itself, starting with cruise control systems and, later, parking assist technologies. Today we see lane-assist technologies, adaptive cruise control and fully automated parking. These developments have combined with innovations coming out of the Cold War associated with global positioning and digital mapping – what today in the UK is termed 'sat nav'. For an excellent overview of future UK possibilities in mobility, we recommend the Rand Europe report *Travel in Britain in 2035* (Rohr *et al.*, 2016). The various developments, combined with the emergence of low-cost sensors, place us on the cusp of achieving the fully autonomous driverless car. The journey to such a technology has been a long and incremental one. In summary, we see technologies converging: first the electric vehicle (EV), referring to the electric power train seen in vehicles such as those produced by Tesla Motors, but also including other concepts such as plug-in hybrids or even hydrogen fuel cell EVs; and second the use of IT in vehicle control as in the 'autonomous vehicle'. We suggest that the two innovations will one day be seen to be two sides of the same coin that will be associated with the most significant evolution in mobility since the advent of the true family car 100 years ago, but the impacts will not simply be restricted to the domain of mobility – they will be seen in energy as well, and most especially in electricity.

2.3. New mobility and the implications for the electricity system

In the UK the electricity distribution company Western Power Distribution (WPD) has been examining the future of power distribution in the face of changing expectations including issues arising from the electrification of mobility. Its study, known as Electric Nation, is uncovering some interesting insights (WPD, 2017). The project is ongoing, but it has already revealed that (in 2018) smartphone apps allowing customer interaction with smart charging systems (i.e. those that manage EV charging while protecting electricity networks from excessive loads) are seen as important to customers – but they must be relatively simple and easy to use. Two app-based systems having been tested.

There are indications that 25% of EV drivers charge their vehicle in the late evening using a timer (Storer, 2018). The Energy Nation team expects that some of these timer users might be accessing the very long-standing UK off-peak tariff known as Economy 7 (originally introduced in 1978 for night time electric storage heaters). If this is occurring it cannot explain the whole 25% figure noted above. Electric Nation notes that there are many customers without a dual tariff (e.g. Economy 7) making use of timers so as to charge at night. It seems likely that this is a result of well-informed users keen to shift their charging to a time of low system demand even in the absence of any incentives on them so to do. As such, these individuals are using timers because of a relatively sophisticated understanding of the system issues coupled with a desire to do the right thing.

UK innovations extend into the retail supply of electricity with the introduction by one small entrant energy supplier, Octopus Energy, of an 'agile tariff', which permits low-cost energy use if congested (early evening peak) times are avoided by consumers. The

Electric Nation project has indications that such incentives are effective with roughly 50% of consumers. At this level of influence the consequences of incentives can be significant and sufficient for effective levels of electricity-demand peak-shifting.

As suggested earlier, UK and other north-western European cities face a particular set of issues shaping the emerging smart city, for example as a consequence of factors as diverse as heritage, weather and latitude. Notwithstanding the successful impact of incentives and the existence of users who spontaneously act so as to reduce pressure on the electricity system, UK EVs have the potential in the future to make the winter peak problems worse in supply and demand matching unless care is taken (Storer, 2018). Unplanned marginal winter additions, such as someone returning home from work and immediately thereafter charging their vehicle, may have to be met through the use of fossil-fuel-fired power generation. The winter evening charge is badly timed but also deeper owing to the use of lighting systems and passenger compartment heating during winter driving (Storer, 2018).

A key technology and policy question going forward is whether consumers will accept charging restrictions (to, for example, specific time of night). It is noteworthy that today in the UK today most EV drivers charge their vehicles at home (Storer, 2018). In addition, the majority of EV drivers in the UK are affluent, and hence are more likely to have off-street parking and to own their own homes, making it easier to fit and use an EV charging point at home.

The Electric Nation project (Figure 2.4) is already seeing that plug-in hybrid EV (PHEV) drivers are the ones that do the deepest charges, although the energy consumed is

Figure 2.4 Eligibility area for the Electric Nation Project. (Source: Electric Nation, 2019)

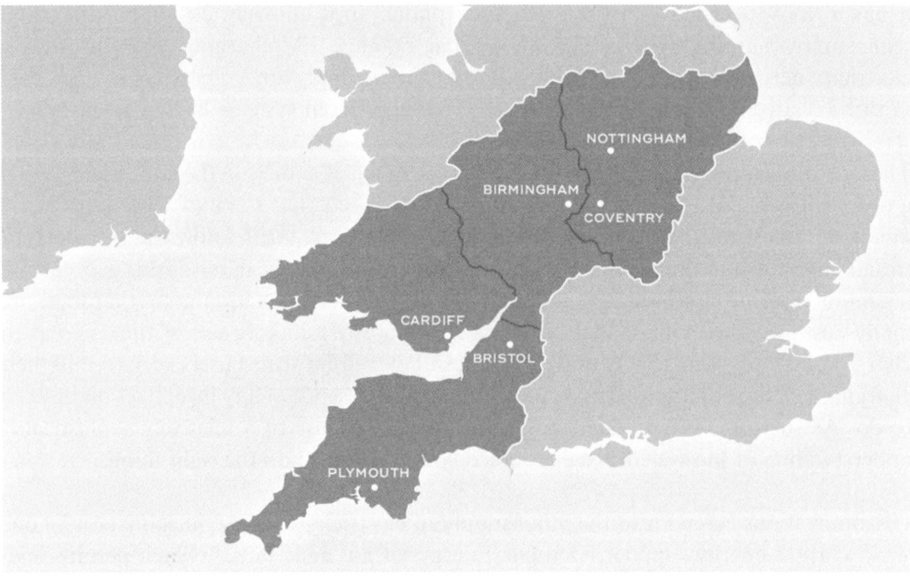

relatively small, as PHEVs typically have smaller batteries than battery EVs (BEVs). This is presumed to be because they are willing to risk a deeper drain of their battery reserve. The fall-back option of petroleum-powered mobility alleviates the range-aversion often encountered among BEV drivers (Storer, 2018).

There has been much consideration of the potential of the stored electrical energy in BEVs and the notion of 'vehicle to grid' power smoothing. The idea has been taken forward by Coomans, an award-winning graduate student from the Netherlands (TU Delft, 2016). J. L.Coomans (2015) has considered the potential for a 'car park as a power station' in which the entity contracting with the electricity system would be the car park and then each motorist would specify their needs to the car park operator (pay to charge, discharge for remuneration or something in between). Coomans has used the techniques of agent-based modelling to study the dynamics of such a proposition. We shall consider agent-based modelling further in Chapter 8.

2.4. The rise of autonomy

While cities are clearly complex technical systems, they are also social constructs. People are central to the future of the city, and such matters are considered in detail in Part 4 of this book. The future of mobility combines electric power trains with powerful IT, including machine learning and artificial intelligence. Such a future brings with it the prospect of autonomous vehicles. These can range from small delivery robots suitable for pizza and similar items, such as those produced by Starship Technologies (Figure 2.5), up to full-sized autonomous vehicles, as discussed in Chapter 5.

Figure 2.5 Starship delivery robot in Redwood City, California. These robots can also be seen in Milton Keynes, UK. (Source: Christopher Down, licensed under CC BY 4.0)

Full autonomy opens up a range of possibilities, including driving party-goers home after a night of drinking to delivering children to school, sparing parents the chore of the school run. But there is more to the potential future than avoidance of the need for a designated driver or a chauffeur-parent. The greatest impacts could arise from the idea that in the future cars will no longer require anyone inside them as they move around the city. In such a situation the city might have a pool of vehicles on call and ready for use that can arrive at a user's residence with a tap on a smartphone app. Such a step builds naturally on the business models developed by Uber and other innovative mobility companies. In such a future city, residents will lose any incentive to own a car of their own. The consequence will be fewer cars in the city, and the visible impact will not be seen in congestion (as the need for mobility will be little affected) but, rather, it will be in a significant reduction in the number of cars parked at the roadside. The cars that are used in the city will be moving for a far greater fraction of the time than at present. This in turn will affect the use of space in the built environment. City side streets will have much wider useful area. These might be set aside for small autonomous vehicles, cyclists and scooter users or could be used as new urban green space.

But always with innovation there are drawbacks as well as benefits. The idea is circulating that 'owners' of driverless cars might set them to drive around town while they are shopping, in order to avoid the need for parking and also to dodge parking charges. Parking, vehicle ownership and congestion are linked issues, and there is the possibility that current regulatory norms and assumptions will be found to be unfit for purpose. For example, relatively recent congestion impact assessments might turn out to be very far from accurate. As planners start to imagine possible futures, they are impeded by a lack of data. We can relatively easily know how many people take a particular bus. We can even track whether they got on a second bus or a train, but what we struggle to know are the actual start and end points of their journey. That is the experience that should be optimised, not the transition from the number 7 bus to the A tram. One London-based smartphone app service is Citymapper. It advises people how to make best use of the capital city's complex system of hired bikes, London buses and underground trains. Citymapper is already using the information it gleans about its users' actual starting points and desired destinations (rather than their entry and exit point from the transport system) to put together new routes (Coldwell, 2017). The company is now launching a unified pay-as-you-go scheme that incorporates its services alongside the traditional services.

2.5. Aligning costs and benefits: a policy challenge

From July 2019 all new home charger units for EVs in the UK must be capable of smart charging (OLEV, 2019). Smart charging has the potential benefit of balancing fluctuating electricity generation (as might arise from a high penetration of intermittent renewable sources) with fluctuating demand (especially short-term diurnal variations). It also has a potential role in addressing another concern – whether the local electricity distribution infrastructure will be adequate to meet local demand in all circumstances.

Consideration of future electricity distribution challenges is pointing to a looming governance issue. Something that somewhat flippantly is becoming known as the 'his and hers Tesla problem'. Imagine for a moment a very affluent neighbourhood in the suburbs of a city such as London. Tesla vehicles are potentially among the most energy hungry of the EV options available to consumers. This is a consequence of these vehicles having among the largest battery packs seen in mainstream EV cars. Tesla as a company is taking measures to mitigate the potential problems caused by its products, one such solution being the Tesla Wall domestic battery pack, but it is entirely imaginable that Tesla owners will not engage with that aspect of the Tesla proposition. Imagine our rich suburban street with many high-power EVs and little by way of load management and demand shifting. Such a situation could easily lead to system demand in excess of the distribution grid's ability to supply. This prompts one of two possible solutions. The first solution would be to implement a smart-charging scheme for the community, effectively rationing the ability to charge during peak hours. In this case, charging is effectively controlled by the distribution network operator. The alternative would be to reinforce the distribution grid locally, so as to meet the expanded demand. Of course, in extreme cases both measures together might be required. The looming socio-technical problem is that smart charging places a restriction that does not currently exist on consumers, and this is likely to impact initially on influential and affluent communities. These consumers might accept the constraints willingly as their contribution to a low-carbon future, but equally it might be regarded as a burdensome imposition. Meanwhile, the latter option of distribution grid reinforcement would represent, in most cases, a socialised response whereby the cost of the reinforcement would fall on all electricity consumers and not just the rich consumers who had arguably caused the problem in the first place. Socialising the cost of smart chargers is a significant economic choice, but socialising the cost of the distribution grid is substantially more significant. In considering such a shift of significant costs away from our rich suburban street to poorer neighbourhoods, we should note that socialised electricity costs are far more regressive than general taxation. As such, the prospect of a deep move towards electric vehicles in affluent suburbs risks being a strong example of a 'reverse Robin Hood' policy, not unlike how the UK feed-in tariffs for domestic renewable energy from a few years ago were subsidies accessible by affluent home-owners but were socialised across all electricity consumers.

These equity and governance issues are a contemporary challenge for policy-makers. Generally, there is emerging evidence that best policy for distribution grid adequacy could see a trend for reinforcement in rural areas and distribution-motivated constraints implemented via smart solutions in urban areas. Such a balance would appear to be the most appropriate response to the socio-technical problem, in part because it combines effectiveness with equity. Factors to be incorporated in such thinking would include: Where are distribution grid stresses most likely to be seen? To what extent are the system stresses related to EVs or to more traditional loads? What is the local potential benefit from smart solutions? And are EV related smart solutions likely to be sufficient or even relevant?

Whatever the role of smartness in meeting future distribution challenges, it must be acknowledged that if a working smart charging system is deployable it will add to the national ability to better match electricity demand with supply as we seek to move towards a low-carbon future.

Meanwhile, another issue is being considered in the UK. Oxford is leading on UK experience for another EV problem: rich people living in terraced town houses. That is, with no easy capacity for off-street charging, how can early-adopter urban EV owners best charge their cars? As noted earlier, perhaps the wider shift to driverless cars will permit an opening of urban space to make public EV charging easier.

There is much media interest in the prospect of charging EVs from street lamps, but this proposition is contested by those who point to serious electrical engineering obstacles. The current carrying capacity of such circuits is insufficient, and, in addition, electrical earth requirements prohibit the use of EV charging equipment too close to street lamps. Metropolitan charging post solutions are possible, but they will not come from a simple conversion of street lighting infrastructure.

Regarding private consumer solutions, Western Power Distribution in the UK is installing three-phase power supplies to homes in anticipation of possible future demand – it calls this ‘superfast electricity’ – seemingly a conscious allusion to the ‘superfast broadband’ marketing campaign from BT (WPD, 2018).

2.6. Conclusions

In this chapter we have seen how different European cities are making different choices as to the future of mobility, but change is underway. The changes are rapid, and predicting the future in this space would appear to be impossible. Some trends are perhaps becoming apparent, however: we are already seeing how the future of mobility and electricity are more closely linked than ever before, and we are seeing the importance of good governance and regulation combined with equity and justice. The linkage of such aspects in local policy-making and technology strategies will be vital if we are to be able to build a sustainable future. Issues of justice and equity inform considerations of governance. We shall return to matters of governance in Part 4 of this book (Chapters 9–11). Such themes will recur throughout this book and be examined in more detail. Whatever the future of energy and mobility – it seems clear that it will be smarter!

Acknowledgements

The author is most grateful to Jonathan Flowers, Nick Storer and anonymous peer reviewers for some important and timely insights into the issues discussed. Those that have assisted do not necessarily agree with the ideas expressed here, and all responsibility for errors or omissions rests with the author alone.

REFERENCES

BBC (2003) Smooth start for congestion charge. *BBC News*, 18 February. http://news.bbc.co.uk/1/hi/england/2770721.stm (accessed 19/06/2019).

Beckett A (2003) Ready Ken? *The Guardian*, 10 February. https://www.theguardian.com/politics/2003/feb/10/london.congestioncharging (accessed 19/06/2019).

Coldwell W (2017) Gett and Citymapper to launch new taxi-bus for London commuters. *The Guardian*, 21 September. https://www.theguardian.com/travel/2017/sep/21/gett-citymapper-launch-new-taxi-bus-london-commuters (accessed 19/06/2019).

Cook J (2018) Electric scooter apps face legal ban on UK roads. *The Telegraph*, 21 June, https://www.telegraph.co.uk/technology/2018/06/21/electric-scooter-apps-face-legal-ban-uk-roads (accessed 19/06/2019).

Coomans JL (2015) *Exploring the Operation of a Car Park Power Plant: Formalising the Operation of a System Innovation with the Actor-Option Framework*. Master's thesis, TU Delft, Delft, The Netherlands. https:/repository.tudelft.nl/islandora/object/uuid:1cf70fdf-e024-47d5-917e-af74cda937e9?collection = education (accessed 19/06/2019).

Electric Nation (2019) Eligibility map. http://www.electricnation.org.uk/eligibility-map/ (accessed 19/06/2019).

EU (2019) Urban Access Regulations in Europe. http://urbanaccessregulations.eu/countries-mainmenu-147/france/paris (accessed 19/06/2019).

Lee K (2018) Scoot alors! Will Paris fall in love with the electric scooter? *The Guardian*, 9 July. https://www.theguardian.com/world/2018/jul/09/electric-scooter-paris-mobile-app-lime-startup (accessed 19/06/2019).

OLEV (Office for Low Emission Vehicles) (2019) *Electric Vehicle Homecharge Scheme Minimum Technical Specification*. OLEV, London, UK. https://www.gov.uk/government/publications/electric-vehicle-homecharge-scheme-minimum-technical-specification (accessed 19/06/2019).

Rohr C, Ecola L, Zmud J *et al.* (2016) *Travel in Britain in 2035. Future scenarios and their implications for technology innovation*. https://www.rand.org/pubs/research_reports/RR1377.html (accessed 19/06/2019).

Storer N (2018) *WPD Balancing Act Conference, Bristol, UK*. (Presentation.)

TfL (Transport for London) (2013) Paying Congestion Charge in shops and petrol stations to end as popularity falls. Press Release, 24 July. https://tfl.gov.uk/info-for/media/press-releases/2013/july/paying-congestion-charge-in-shops-and-petrol-stations-to-end-as-popularity-falls (accessed 19/06/2019).

TfL (2015) Discounts and exemptions. https://web.archive.org/web/20150418173557/http://www.tfl.gov.uk/modes/driving/congestion-charge/discounts-and-exemptions?intcmp = 2133 (accessed 19/06/2019).

TfL (2018) *Transport in London, Report 11*. TfL, London, UK. http://content.tfl.gov.uk/travel-in-london-report-11.pdf (accessed 19/06/2019).

TU Delft (2016) Q-Park thesis award for Jurriaan Coomans. https://www.tudelft.nl/en/2016/tbm/q-park-thesis-award-for-jurriaan-coomans (accessed 19/06/2019).

WPD (Western Power Distribution) (2017) *Customer Research and Trial Update Report: Electric Nation*. WPD, Bristol, UK. https://www.westernpower.co.uk/docs/Innovation/Current-projects/CarConnect/Electric-Nation-Customer-Research-and-Trial-Update.aspx (accessed 19/06/2019).

WPD (2018) Superfast electricity for electric cars. https://www.westernpower.co.uk/news-and-events/latest-news/superfast-electricity-for-electric-cars (accessed 19/06/2019).

ice Publishing

Nuttall, William J, Gibson, David V, Trzmielak, Dariusz and Ibarra-Yunez, Alejandro
ISBN 978-0-7277-6425-6
https://doi.org/10.1680/emsc.64256.031

Chapter 3
Smart cities: locations for environmental entrepreneurship

Stuart Parris
The Open University, Milton Keynes, UK

3.1. Introduction

In Europe, smart cities are viewed as an opportunity to stimulate innovative activity either directly or indirectly achieving Europe 2020 policy targets, such as those related to energy usage, carbon emission reduction and improved mobility for citizens (Manville *et al.*, 2014; SETIS, 2019a). To achieve such targets requires innovation and, subsequently, appropriate forms of enterprise and social enterprise delivering to consumers and citizens. Even when existing technology may exist in order to achieve targets, the diffusion of such technology can require substantial engagement with users directly, to help develop the most appropriate forms of environmental products and services for their needs. This often requires an entrepreneurial private sector incentivised to deliver economic and environmentally sustainable products and services, working in conjunction with other public stakeholders. It raises the question of whether smart cities are an appropriate institutional 'vehicle' to achieve these goals.

Entrepreneurs – defined as those who recognise and seize upon opportunities – have often been viewed as the very actors responsible for creating some of the most serious environmental challenges faced at both the local and global levels (Lenox and York, 2011). Business actors, pursuing self-interest, when set against a world with limited natural resources, can damage the environment through their activities, generating negative externalities, particularly without necessary correction and regulation.

Those defending entrepreneurship, however, point to a phenomenal rise in technical capability in recent decades and the potential of technology to reduce pollution and carbon emissions. Debate over the potential of technology to overcome limits to growth have a long history dating back to Malthus. Such debates were reinvigorated by the report of Meadows *et al.* (1972) to the Club of Rome, noting that new innovation can support sustainable economic growth, but also exacerbate environmental problems if left to market forces alone. Interest in the potential of environmental entrepreneurship has steadily grown and been investigated as a means for supporting sustainable economic growth (Lenox and York, 2011). Yet concerns remain about smart city projects representing utopian visions of technology-led development without acknowledging the limitations of entrepreneurship to tackle complex social and environmental challenges (Kummitha, 2018).

In parallel to debates over the role of technology for society and environment, a substantial literature on economics, urban planning and innovation has grown around a critique of mainstream economics, instead focusing on the dynamics processes at play in the development of innovation within and between firms (Nelson and Winter, 1982). This literature has focused on the quality and function of national innovation systems to support the development of innovative projects to bring together different actors, and resources, within a shared vision between government, industry and research communities (Foxon *et al.*, 2005). It has long been recognised, however, that innovative activity tends to be concentrated geographically, often around key cities, such that localised regional innovation systems (involving public and private actors) have an important role to play in building innovative capacity, institutionalised learning and productive culture (Cooke, 2001; Cooke, 2007). The smart city, a more recent and related approach, adopted at a city level, has also integrated a focus on the concept of 'smart' urban growth to deal with the challenges of increasing urbanisation, population growth and the efficient use of resources, where 'smart' is centres on the role of forms of development driven by information and communications technology (ICT) (Caragliu *et al.*, 2011). As with innovation systems approaches, research has indicated the crucial and interrelated role of human capital, innovation, entrepreneurship and skilled labour in providing the foundation for 'smartness' and urban development (Berry and Glaeser, 2005). A key uniting feature between innovation systems literature and smart cities is the role attributed to social and relationship capital, based on the need to collaborate, learn and adapt to new innovation and technology, and so necessitates local absorptive capacity for knowledge spillovers (Caragliu *et al.*, 2011).

Although the key driver for smart cities has been focused on the role of ICT, networked infrastructure and connectivity to deliver smart urban growth, increasingly smart cities have focused on social and environmental sustainability, giving rise to a variety of definitions. One popular model of smart cities has been to evaluate them based on a combination of, and relationships between, six characteristics, including smart economy, smart mobility, smart environment, smart governance, smart living and smart people (see Chapter 1 and Centre of Regional Science, Vienna University, 2007).

In line with smart city characteristics, this book is structured around key themes in smart cities. This chapter and the next complete Part 1 of the book, giving an overview of smart city characteristics. Subsequent parts of the book delve into the mobility, energy and governance characteristics of smart cities in greater detail.

The objective in this chapter is to explore the potential link between cities with smart environment initiatives and the location and characteristics of innovative activity, as well as the characteristics of firms operating in the wider environmental sector and how they relate to smart cities. To achieve this objective the chapter considers how smart cities can support innovation and entrepreneurship by working with related literatures. Part 2 of this briefly considers the smart city construct and draws out relationships to the literature on environmental entrepreneurship. In Part 3 this chapter is developed further by examining the theoretical and empirical literature on what makes an innovative location. Part 4 uses factors from the literature associated with environmental entrepreneurship and innovative locations, working with empirical data to assess the innovative and entrepreneurial potential of smart city locations. Part 5 considers networks and social capital as features identified as crucial to both entrepreneurship and innovation, and necessary to bridge between all city stakeholders to improve outcomes. Part 6 concludes the book.

3.2. Smart cities defined and their purpose

The smart city concept is both fuzzy and contested (Caragliu *et al.*, 2011; Hollands, 2008; Richter *et al.*, 2015). There is no single definition of what is required for smart city status, and many cities across the world have adopted aspects of smart city activity, or at least have plans to do so. In Europe the smart city initiative has been heavily supported by the EU, the European Commission and related directorates and agencies. Funding from European agencies has been made available to support cities to deliver on the smart city agenda, and also continue work developed under previous guises and programmes, such as ‘future cities’ and ‘sustainability cities’, to name two. An example was the Intelligent Energy Europe programme (IEE), launched in 2003 to deliver energy efficiency, renewable energy and green mobility. It mainly focused on large urban cities, with the idea of encouraging European industry to establish leadership in efficiency and clean technology. As this overlaps with the smart city agenda, it became part of (and overlapped with) the European Strategic Energy Technologies Information System initiative on smart cities (SETIS, 2019b) (Figure 3.1). Part of European thinking on smart cities, was to – as in the IEE programme – challenge global competition in the development of new technology areas as well as deliver sustainability goals. This highlights the fuzziness of the smart city concept, as one aiming to deliver on multiple fronts, and also how challenging global competition to achieve global industrial leadership in technology areas can in itself be a direct challenge the goals of achieving sustainability (Lazaroiu and Roscia, 2012).

Furthermore, according to a European Parliament report evaluating 20 European smart cities, the majority had initiatives focused on European Commission smart sustainable and inclusive growth targets and 25% explicitly targeted employment (Manville *et al.*, 2014). Although technological innovations are regarded as having the potential to contribute solutions to mitigate the environmental impacts of economic activity, few smart cities were found to have a specific focus on research and development (R&D) and related innovation that would help foster private sector innovation and additional R&D spending (Manville *et al.*, 2014). This highlights the challenge of achieving innovation and economic leadership.

Richter *et al.* (2015) identified seven definitions of a smart city from sources predominantly in regional and urba development literature, as well as economics, engineering, and an early definition from the Brookhaven National Laboratory in 2000. They then extracted the most commonly applied elements to produce their own definition, from those definitions already in use.

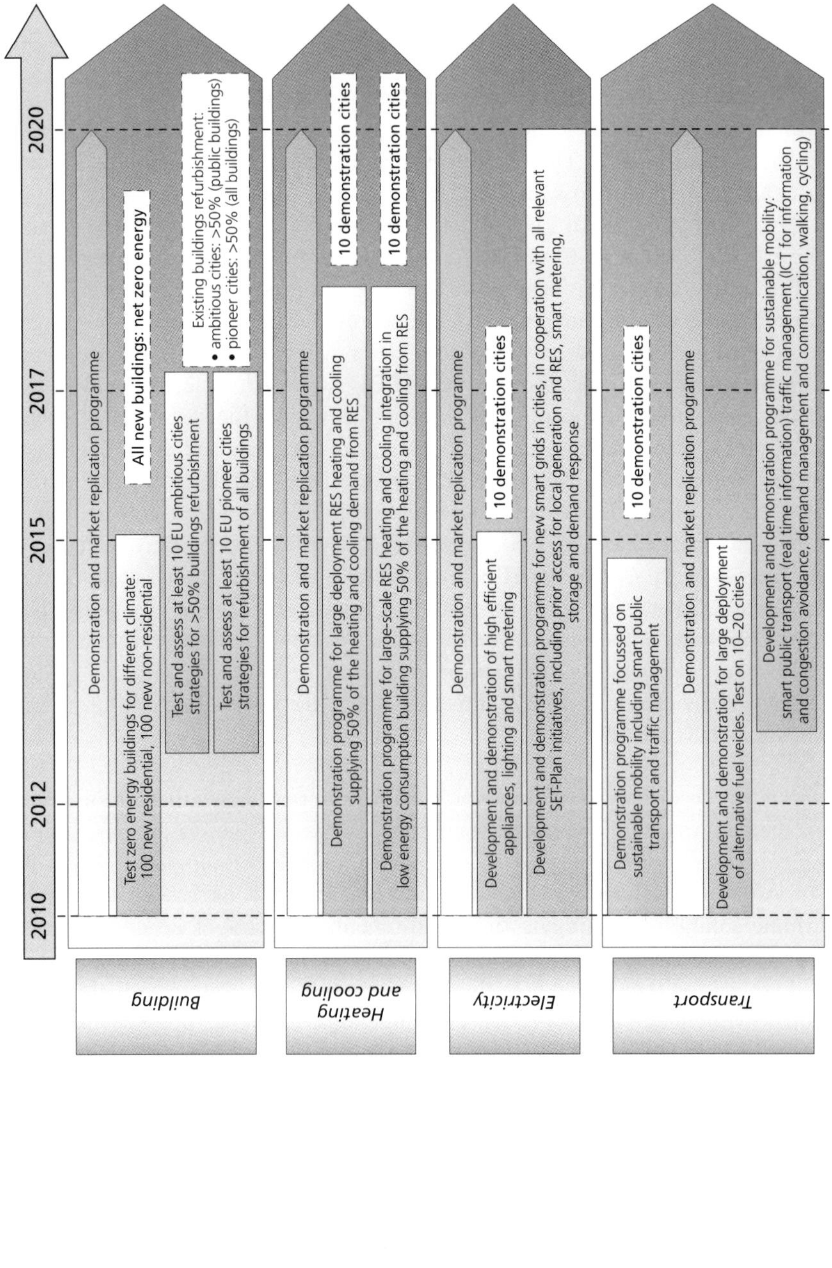

Figure 3.1 SETIS smart cities indicative road map (SETIS, 2019b). RES, renewable energy sources; SET-Plan, European Strategic Energy Technology Plan

> A Smart City is an agglomerated area affected by a high concentration of learning and innovation as a result of creative citizens and institutions as well as the implementation of a digital infrastructure with the overall objective of achieving economic growth and a high quality of life, while keeping in mind the scarcity of natural resources.
>
> (Richter *et al.*, 2015, p. 214).

In an attempt to synthesise the approaches taken towards understanding smart cities, Richter *et al.* (2015) highlighted six characteristics of smart cities, shown in Figure 3.2. ICT is frequently considered the most important characteristic, providing the essential communication and infrastructure needed for 'smart' activity, including the provision of information via digital systems to allow systems to be designed efficiently for consumers and citizens and a platform for innovators to work from, as well as effective city-wide resource management, distribution and recycling, such as sensor systems, cloud technology, energy management and monitoring via smart grids.

Figure 3.2 Six characteristics of smart cities. (Adapted from Richter *et al.*, 2015)

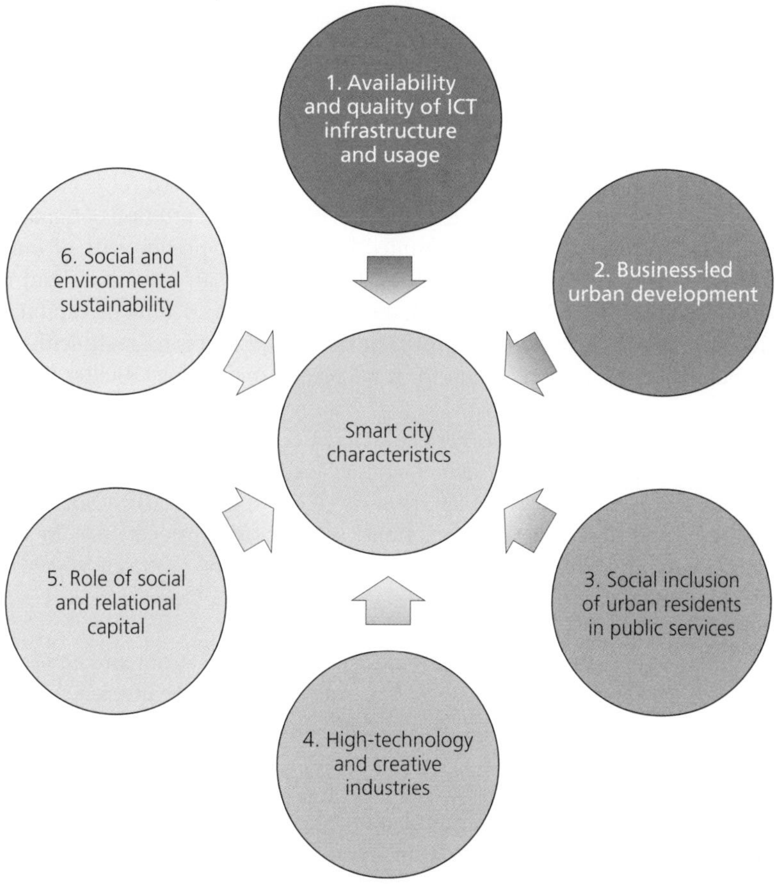

Secondly, business is a function of the vast majority of cities in the world. Even with public support, smart city activity requires extensive engagement of business to deliver the full range of products and services that citizens demand, and public investment can often only seed fund a city's progress towards becoming a smart city (Richter *et al.*, 2015). However, for advances to be made and city activity transformed, entrepreneurship is increasingly necessary to provide a flow of new ideas, innovation and firm creation such that, increasingly, innovative businesses are seen working in conjunction with public organisations within smart cities (Lombardi *et al.*, 2012).

Thirdly, the inclusion of residents and their access to public services has been identified as part of smart city activity by Richter *et al.* (2015). In particular, smart cities help citizens access the services they require more efficiently. This characteristic, as stated by Richter *et al.* (2015), has been broadened further by some, noting the vital role of citizen engagement in successful smart cities, therefore being not simply about access to services but also about shaping services and the city itself (Akande *et al.*, 2019). A 2008 Siemens report on London, for example, highlighted that success for sustainability in the UK's capital city would require extensive cooperation, particularly with consumers who could, via their purchasing decisions, achieve approximately 70% of the maximum potential carbon dioxide abatement (Siemens, 2008). Therefore, engagement with a city's citizens, and allowing them access and representation in discussions, is key.

The fourth characteristic is growth from high-technology and creative industry arising from the contribution of human capital, the skills and knowledge applied to solve problems and support creative endeavours. This is separate from the 'hard' city infrastructure, although with a significant interdependence with it: it is the concentration of human capital that, according to Florida (2002), has resulted in the greater productivity of knowledge workers in city locations. ICT infrastructure, for example, can help share and manage information and codified knowledge, but the concentration of human capital activity within cities allows for the development of tacit knowledge that is more difficult to codify and store, but is nevertheless important for learning and innovation (Richter *et al.*, 2015).

Fifth, and again closely related to other characteristics, there is the role of social capital. Ties fostered between individuals help facilitate learning – knowledge exchange that contributes to the overall success of smart cities (Kourtit *et al.*, 2012). Social capital is 'the capital captured through social relations', which are a 'social asset by virtue of actor's connection and access to resources in the network or group of which they are members' (Lin, 2002, p. 19).

Sixth, and importantly, the smart city should have a focus on social and environmental sustainability, providing the impetus to reduce waste, increase the efficiency of resource usage, and promote better wellbeing for citizens, as discussed in Chapter 4.

This section has highlighted the potential fit of entrepreneurship, innovation within smart cities, and the underlying environmental and sustainability goals smart cities seek to achieve. It takes us to the question, 'Can smart cities be a location for environmental entrepreneurship?'

3.2.1 Key ingredients of environmental entrepreneurship

Environmental entrepreneurship, as a concept, covers a range of different themes. At one end of the thematic spectrum lies mainstream entrepreneurship concerning the commercial, for profit, development of new technologies, another theme explores technology to mitigate the impacts of our economies, and at the other thematic extreme are those activities that seek to challenge traditional economic models of development and the traditional use of natural resources, seeking the proximate goal of economic and social sustainability, rather than the traditional goal of profit.

It is also increasingly acknowledged that the whole of the environmental sector needs to deliver substantial innovation to avoid increasing energy costs and to comply with internationally set environmental targets (Bosetti *et al.*, 2009). An awareness of the need for new sustainable and environmental technologies has grown since the 1970s (Shrivastava, 1995) but investments in developing clean and environmental innovations have been limited. Innovation in areas such as clean technology has gained momentum since the 1980s, with the growing political emphasis on sustainable economic development. As an example, the environmental sector features a low R&D intensity profile in the UK and across Europe despite continuously being highlighted by the media and policy-makers as the domain of the most recent technology revolution – the 'green revolution'. The sector accounted for only 3% of the UK's R&D budget in 2010 (Eurostat, 2012), and the energy industry, in particular, displays very low levels of innovative activity, with R&D intensity levels of the oil and gas industry and the electricity industry of around 0.3% and 0.6%, respectively (Hernández *et al.*, 2008; Margolis and Kammen, 1999). Chapter 7 discusses the economics of renewable energy innovation.

Previous evidence suggests that technologies in the renewable energy sector, namely bioenergy, wind power, solar thermal energy and solar photovoltaic energy, have gained acceptance, with double-digit growth in some energy sectors, even though the process of changing the overall energy system has generally been slow and uncertain (Jacobsson and Johnson, 2000). Recent decades have witnessed an increasing number of small and new technology-based firms operating in the environmental sector, spanning a range of environmental technologies. In solar power, innovators have targeted improvements to photovoltaic efficiency, manufacturing cost, light-tracking and energy storage. In wind power, innovators have targeted the development of new alloys and composites to improve turbine performance and longevity as well as technologies to maximise the efficiency of transfer to the electricity grid (Cullen, 2009). Often those firms entering the market are delivering quick and efficient low-energy solutions to the economy (Cullen, 2009; Lee *et al.*, 2009). In particular, small and medium-sized enterprises (SMEs) have increased their share of innovative activities, as larger organisations have outsourced R&D to SMEs (Jacquier-Roux and Bourgeois, 2002). However, the development of innovation in the environmental sector relies heavily on financial investment, and capital shortages can have detrimental effects on the innovative performance of SMEs in high-technology sectors and on the demand for innovative technologies (Madrid-Guijarro *et al.*, 2009). As an example, De Jager *et al.* (2011) estimated an approximate capital expenditure gap of €35 billion per year required across the EU member states to provide the necessary investment in renewable energy sources to achieve the 2020 renewable energy generation targets. Governments across the world have introduced green stimulus plans to boost the demand for environmental technologies in an attempt to encourage firms to invest (e.g. European Commission, 2011), and these policies are also intended to stimulate the private investor appetite for environmental technologies, especially during the recent negative macroeconomic climate surrounding many major economies (Bürer and Wüstenhagen, 2009). Such policies need to take account of both incentives for R&D, promoted by encouraging learning from research, as well as market adoption, promoted via learning by doing (Jamasb *et al.*, 2008). On the supply side, innovative firms, particularly SMEs, require significant resources to fund their innovations, especially at early stages of their life cycle, characterised by low or negative levels of cash flow and high R&D costs (Mason and Harrison, 2004). Indeed, Garnsey *et al.* (2006) reported that access to funds to finance R&D is the major concern for entrepreneurial innovators in the environmental field.

Likewise, uncertainty about the potential demand for innovative environmental products and services, compared with traditional goods, can deter investment. Investors are often concerned that the higher price of environmental goods and services puts customers off. Investors can also be deterred by the costs of creating new markets for environmental goods and services, as this can involve the need for early investment in consumer awareness of environmental issues. Barriers to traditional finance mean entrepreneurs can end up self-financing; for example, Eventure Media (2009) in its Cleantech Survey noted that 80% of the clean-technology SMEs in their survey were self-funded or family/friend/business-angel funded. This can create a need for public investment to help provide entrepreneurs with a bridge to private investment, such that public investment can be used to help deliver proof of concept, or trials of new innovative technology.

While finance is a concern for those entrepreneurs either planning to pursue entrepreneurial activity or in the first few stages of doing so, from a city perspective it is vital to have a supportive, innovative environment. By this we mean an environment capable of generating innovative opportunities as well as able to identify and act upon opportunities; for this it is necessary to have an innovation system.

3.3. Smart cities as innovative centres

Traditionally, innovation and innovation systems have been considered at a national level, with national policy targeting factors to improve innovative performance. However, increasingly, policy has been targeted at the regional level, in recognition of the role of key global locations in supporting innovative activity and a desire to replicate the success of areas such as Silicon Valley. Silicon Valley is a very well-known and oft-cited example of where public investment in basic research and development and a culture of entrepreneurial start-ups have been supported by the availability of risk-based finance. Silicon Valley is associated with many start-ups that have grown rapidly to become much-admired locations for innovation, and it persistently remains a top destination for innovative business.

The smart city, as Richter *et al.* (2015) defined it, shares similar characteristics with locations renowned for innovation and entrepreneurship. The principles understood to support innovative centres, such as high-technology clusters or an innovative system, rely on a co-location of resources, such as human, financial and social capital. Although such places are also shaped by public policy, and in many cases supported by public funding and investment (directly and indirectly), the additional drive for environmental and social sustainability in smart cities is anticipated to have direct and local impact on citizens' lives. This feature of smart cities, with a clearly identified goal and performance judged on sustainability, is a clear difference from other innovative clusters – often judged by measures of commercial performance and economic growth. Nevertheless, as Florida (2002) has observed, the most successful innovative clusters have been co-located with a creative sector, providing an attractive environment for those choosing to support the underlying innovation system. Next, we briefly review some key concepts in the literature on innovative locations.

Like the smart city concept, the definition of a cluster is not universally accepted within the literature (Morosini, 2004), although the general drivers for firms to cluster outlined by Marshall (2009) provide a common starting point. Marshall made three observations in an attempt to understand why companies continued to locate in the same geographical areas. Marshall's first point was that the geographical closeness of companies generated mutual benefit in accessing a pool of specialised talent or human capital, where co-location created a skilled workforce that could meet the needs of an industry. Marshall also suggested that this close geographical location could deliver economies of scale in developing and utilising localised technology innovations or infrastructures, such as roads, termed 'non-traded inputs' (Morosini, 2004, p. 307). Finally, there was also a mutual benefit obtained from the flow of ideas or information, which could easily be exchanged when companies located in close proximity, as opposed to those in geographically dispersed locations.

Porter (1990, 2000) extended and updated Marshall's observations to include a range of additional inputs such as the relationships between universities and business, and management of the supply chain. Porter emphasised the mutual benefit of firms based in the same location competing against one another, but also cooperating and sharing the benefits of specialist suppliers, service providers and trade associations supported by the concentration of activity within a cluster (Porter, 2000).

Although identifying the flexibility of the cluster concept to cover regional, national or international geography, Porter (2000) argued that globalisation has reduced the locational benefits resulting from Marshall's first and second points. In other words, he suggested that knowledge flows were key – such that the development of local specialisations and social networks can create vital sources of competitive advantage for firms, particularly in the internet age. In this respect, clusters can vary according to the nature of innovation processes and structural conditions such that concentrations of firms evolve (Iammarino and McCann, 2006). Places such as Silicon Valley grew from a localised cluster formed around strong local ties and relationships to a large agglomeration of competing firms, supporting the central role of networks in innovation (Iammarino and McCann, 2006; Lundvall and Borrás, 1997). As Lundvall and Borrás (1997, p. 104) noted, 'Only a small minority of firms and organisations innovate alone, and ... most innovations involve a multitude of organisations'.

The use of networks, and the activity of networking to establish links between people, teams or organisations, is a fundamental factor in successful innovation projects. Studies of innovation have demonstrated the importance of formal and informal networks. Research suggests informal networks are very important, with in-house R&D teams benefiting from collaboration and stimulation from outside their respective organisations (Freeman, 1991). Social networks, as an 'organisational structure', are seen as ways to improve the effectiveness of innovation, increase the innovative performance of network actors and, if successfully operated, provide a source of innovation (Van Aalst, 2003).

Social networks are an effective way to share a variety of sources of information; if used effectively they can provide open access to information from a wide range of people, cover wide geographies, and reduce hierarchical or organisational boundaries. In essence, networks allow up-to-date information to move freely between organisations and individuals. In this way, networks can help facilitate the sharing of information that might otherwise be challenging due to barriers within and between organisations and market structures.

In particular, the role of social networks in the development of innovation has been given emphasis, with researchers highlighting a role of networks to help facilitate the sharing of tacit knowledge, often vital for innovation. Tacit knowledge, in this context, is taken to mean informal knowledge, difficult to codify in journals and books, including 'the knowledge of techniques, methods and designs that work in certain ways and with certain consequences, even when one cannot explain why' (Rosenberg and Nathan, 1982, p. 143). Senker (1995, p. 444) observed 'that tacit knowledge can be acquired only by personal experience or personal interaction with experts'. The difficulty of transferring tacit knowledges means it is sometimes referred to as 'sticky' knowledge.

Tacit knowledge is also important for the interpretation of formal codified knowledge. Often, tacit knowledge provides the necessary understanding of the context and process of how formal knowledge was developed and how best to interpret or exploit it further. Social networks, therefore, are an important means for people to share tacit knowledge, to make sense of other formal knowledge so that it becomes more effective in application. The effectiveness of networks for sharing and learning relevant knowledge may also accelerate the speed of such knowledge transfer, reducing the need to codify certain knowledge in published books or reports (Lundvall, 2000; Van Aalst, 2003). As innovation has become increasingly open and distributed over time, coupled with a greater need for collaboration and use of outsourcing, the innovation process has become increasingly complex. Social network activity, particularly when organised within innovative clusters, is recognised as conducive for tacit knowledge, contributing to the innovative performance of such locations (Iammarino and McCann, 2006; Storper, 1993).

It is the 'sticky' tacit nature of knowledge that leads to the view that institutions are at the centre of the development of new knowledge: for example, research institutes and universities would be central to the development of concentration of innovative activity. The triple-helix model adds universities to private firms and public administrations as a third and central force in the innovation system (Etzkowitz and Leydesdorff, 2000). Such organisations can provide a source of knowledge, encouraging clusters of firms to gain access and engage with new knowledge, and also to access the related supply of skilled labour (Mytelka, 2004); secondly, they can provide a source of innovation or knowledge with commercial potential. They can be active in creating the necessary social conditions in the local environment through interaction with and support of local innovative business either directly by becoming academic founders, as an activity often encouraged by universities, or indirectly supporting the development of the cluster through local relationships with business (Keeble *et al.*, 1999; Leibovitz, 2004; Stuart and Ding, 2006). Such ties are further strengthened as the process from academia to firm is rarely linear, as both academics and academia benefit from external knowledge and also income from engagement and knowledge exchange with non-academic actors (Bower, 2003; Probert *et al.*, 2013). Although the importance of localised inter-firm networking finds less empirical support, existing companies can also provide a source of new innovative activity and some anchorage of new enterprise to the parent location, as new entrepreneurs leverage existing ties.

Crucially, however, only certain regions evolve to support a concentration of high-technology business (Cooke, 2004). Feldman and Francis (2003) studied the Capitol Region, and found that in pursuing their own individual interests, entrepreneurs acted as the 'critical element in the formation and the vibrancy of clusters of technology intensive firms' (Feldman and Francis, 2003, p. 780) by drawing together the necessary resources to support entrepreneurial firms in the region. This was also found by Von Burg and Kenney (2000) in their research on venture capital, who concluded that investment must be actively recruited by the firm; there are no guarantees that any particular firm will obtain finance, and the crucial difference is the entrepreneur's ability to find the resources needed. In this way, a particular region can develop a regional innovation system as a product of routines relating to a particular technology or sector, producing a concentration of related organisations (Niosi and Banik, 2005). This wider view of the processes occurring within concentrations of firms has similarities with a description of a regional innovation system (RIS) defined as

> geographical concentrations of interacting organisations (innovative firms, research universities, government laboratories and venture capital firms) aimed at the development of specific technology.
>
> (Niosi and Banik, 2005, p. 343)

Our discussion to this point may suggest that concentrations of high-technology activity will only arise in regions that are able to learn how to develop the interactions of local entrepreneurs, firms and institutions. This involves a cumulative learning process, including the continual 'integration of up-to-date tacit knowledge into innovation processes', to maintain or further develop their 'competitive edge' (Cooke, 2004, p. 175). Naturally, this does not preclude engaging with geographically distant collaborators, suppliers or customers. Important knowledge and information may be sourced from outside the RIS (Niosi and Banik, 2005): for example, large companies of strategic importance may be located outside the region in focus; equally, firms located within particular clusters may have strong national or international relationships, particularly when it comes to accessing cutting-edge specialist R&D knowledge (Cantwell and Iammarino, 2000; Mytelka, 2004).

Our next step is to investigate whether the underlying dynamics of innovation can be related to smart cities empirically.

3.4. Smart cities as a location for environmental entrepreneurship

From a theoretical basis there are many reasons to expect that those locations with exciting innovation and entrepreneurial systems can also be smart cities. Using a dataset from a report produced for the European Parliament mapping smart cities in the EU (Manville *et al.*, 2014), we looked to understand relationships between places with high levels of innovative and entrepreneurial activity with the locations of smart cities. The data included 468 cities with a population greater than 100 000 collected in 2013–2014. Data was also provided on each city's progress towards becoming a smart city, identifying and coding specific policy according to smart categories, such as smart economy, smart people, smart environment, smart governance, smart living and smart mobility, as well as the general maturity of smart city initiatives.

Of the 468 cities included, 240 had a smart city policy, with the most frequent initiative being a smart environment (199), followed by smart mobility (125). Only 67 cities had a smart economy initiative.

The list of 468 cities was matched with Eurostat NUTS 2/3 data, to provide city, municipal or regional measures of key activity, to examine the relationship between the location of smart city initiatives and city/regional entrepreneurial and innovative capabilities.

Additional data selected from Eurostat included measures of factors identified as important for regional systems of innovation, including entrepreneurial activity from the Regional Entrepreneurship Development Index (REDI). Two key measures of entrepreneurial activity were identified, the first an overall measure of entrepreneurialism (including individual level and institutional level measures), the second an index based on the entrepreneurialism of individuals. The other measures were

- innovative activity (patenting activity)
- human capital (high-level degrees)
- R&D environment (government, business and higher-education investment in R&D)
- ICT connectivity (level of broadband provision)
- regional business activity (composition of manufacturing, services and technology in a region).

Each measure was scaled by the size of the underlying population in each city region. To account for differences in general economic performance, the log of the city population was included, and also disposable income as a measure of wealth. Wealth has been found to be linked to the likelihood of cities adopting smart initiatives (Akande *et al.*, 2019).

Using a logistic regression analysis we regressed a measure of smart city maturity against factors capturing entrepreneurship and innovation within the same locations. We also conducted a probit regression analysis of the same factors on the likelihood of a city having a smart environment or smart economy policy in place. The results are summarised in Table 3.1, with the statistical strength of the relationship and the direction of the effect indicated (+ refers to a positive influence of the factor on the probability that a smart city is more mature, or having a smart economy or smart environment policy). It should be noted that the results are based on cross-sectional data, and should be interpreted as indicative of a relationship between the presence of particular city region characteristics and smart city activity, rather than a causal relationship.

Table 3.1 highlights an interesting observation that our measures of entrepreneurial activity were not related to the maturity of smart cities, nor smart economy policy. The measure of entrepreneurial activity of individuals within a city region had a positive association with cities in the sample having a smart environment policy. When institutional-level entrepreneurial activity was included, the relationship was only statistically significant with smart mobility activity, but the relationship was negative.

Table 3.1 also highlights interesting differences between the relationship of traditional measures of innovative activity and smart cities. For example, a higher concentration of graduates and high-technology manufacturing made it less likely that a city will have a smart environment or smart mobility policy and less likely to be a mature smart city. Curiously, those attributes traditionally associated with innovative clusters were not associated with smart cities.

In contrast, the greater concentration of academic research activity within a city was associated with smart cities that are mature and have a smart environment policy, supporting the role of universities within smart cities. It also appeared that locations with a concentration of industrial manufacturing activity, rather than high-technology activity, were also more likely to be associated with smart environment policy. This perhaps indicates that those cities with a legacy of industrial manufacturing and the related environmental challenges are responding directly to these problems by embracing the smart city.

Those cities with smart mobility policy showed a positive association with business R&D, rather than academic R&D, which represents a different knowledge base, perhaps signalling a more effective transfer of applied research or learning by doing, rather than benefits from learning from academic research.

The results in Table 3.1 are exploratory, and further research is needed on the long-term dynamics of smart city activity. Taking the results of Table 3.1 overall, this may indicate that smart cities are emerging in response to local legacy issues. They may be developing in locations without a history of high-technology development, nor large local graduate

Table 3.1 Estimations of smart city factors

Measure		Specification							
		Smart maturity		Smart economy		Smart environment		Smart mobility	
		1	2	3	4	5	6	7	8
Entrepreneurship index	REDI score (institutional and individual components)	NS		NS		NS		(–)**	
	REDI (individual component)		NS		NS		(+)**		NS
Human capital	Tertiary education	NS	NS	(–)*	(–)**	(–)*			(–)**
Regional economy	High-technology employment	NS	NS	(+)***	(+)**	NS	NS	(+)*	NS
	Manufacturing	(+)**	(+)***	NS	NS	(+)**	(+)**	NS	NS
	High/medium-technology manufacturing employment	(–)***	(–)***	NS	NS	(–)***	(–)***	(–)**	(–)**
	Professional employment	(–)*	NS	NS	NS	(–)*	(–)**	(–)**	NS
	Administrative employment	(–)**	(–)*	NS	NS	(–)*	(–)*	NS	NS
Innovation	Patent applications	NS	NS	NS	NS	NS	NS	NS	(–)*
Connectivity	Broadband	NS	NS	NS	NS	NS	NS	NS	NS
Research environment	Business R&D	(+)***	(+)***	NS	NS	(+)**	NS	(+)***	(+)***
	Government R&D	NS	(+)*	NS	NS	NS	NS	NS	NS
	Higher-education R&D	(+)***	(+)***	NS	NS	(+)***	(+)***	(+)*	1NS
City scale	log(population)	(+)***	(+)***	(+)***	(+)***	(+)**	(+)***	(+)***	(+)***
Wealth	Disposable income	(+)***	(+)*	NS	NS	(+)**	(+)**	(+)***	(+)***

***, significant at 1%; **, significant at 5%; *, significant at 1%; NS, not statistically significant.

populations relative to the total population; such skillsets, seemingly important to smart cities, might be shared from elsewhere. The smart city initiative includes a broad range of activity, thus it might be unrealistic to expect a concentration of co-located innovative activity within smart cities.

Table 3.1 also indicates the clear difference between factors related to smart economy and smart environment policy. In particular, the presence of a smart economy policy in a city has very little association with either innovative or entrepreneurial activity (as measured here). It demonstrates that the breadth of the smart city is a broad concept – addressing all characteristics in tandem is challenging, as not all aspects of 'smart' necessarily link with environmental sustainability.

The smart city, as a relatively new concept, has been described by some researchers as demonstrating a weaker association with environmental issues, compared with the sustainable city concept (Gargiulo Morelli *et al.*, 2013). Viitanen and Kingston (2014) have highlighted the role of smart cities in displacing environmental challenges to disadvantaged locations or social groups, rather than solving them. De Jong *et al.* (2015) also observed, via a bibliometric study of the sustainable urbanisation literature, a clear separation in the use of 'smart' versus 'sustainable' for cities. Their work highlights that environmental issues remain core to the discussion of the sustainable city concept, while the smart city concept is linked to engineering solutions and investment opportunities, shifting the focus somewhat from environmental issues to one of infrastructure development and commercial activity. Similarly, as summarised in Figures 3.3 and 3.4, Ahvenniemi *et al.* (2017) found, in their comparison of eight smart city and eight sustainable city policy frameworks, that although smart city frameworks include a clear focus on social and economic aspects of city life, their design gives less significance to environmental issues in cities. In contrast, however, Ahvenniemi *et al.* (2017) found city sustainability frameworks to have a more even focus on social and environmental aspects, but less on the economic dimension.

Furthermore, Figure 3.4 shows the clear difference between the smart city and sustainability frameworks in terms of the coverage of different functions of cities. As seen in Figure 3.4, smart city and sustainability frameworks only partially overlap, and therefore a distinct challenge of combining smart and sustainable initiatives remains at the city level. Similarly, the lack of integration of energy, environment and transport within the smart city frameworks found in Ahvenniemi *et al.* (2017) offers a partial explanation of our results in Table 3.1.

In response to the challenges of combining 'smart' and 'sustainable' initiatives at the city level, Akande *et al.* (2019) have developed a smart and sustainable index of European cities, to give greater prominence to the sustainable aspects of smart cities. This index highlights in particular the key differences between European cities and their diversity of performance (Figure 3.5).

The Akande *et al.* (2019) index shows the large divide between European countries with leading cities, according to their index, in northern Europe (Figure 3.6). Akande *et al.* (2019) noted that cities scoring highly in their smart sustainable index were within close

Figure 3.3 Relative occurrence of the number of indicators for both smart city and sustainability frameworks with respect to the three dimensions of sustainability. (Source: Ahvenniemi *et al.*, 2017)

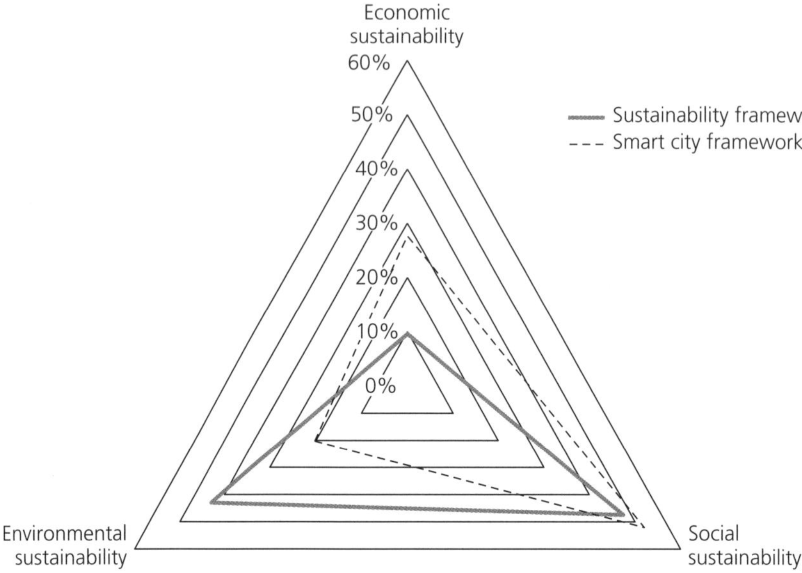

geographical proximity, and predominantly northern European cities with a high GDP per capita. The authors suggested wealthier cities have greater freedom to focus on issues such as sustainability, as well as the potential for neighbouring cities to learn from each other and exchange ideas.

Figure 3.4 Division of the indicators of both smart city and sustainability frameworks under the ten sector categories. (Source: Ahvenniemi *et al.*, 2017)

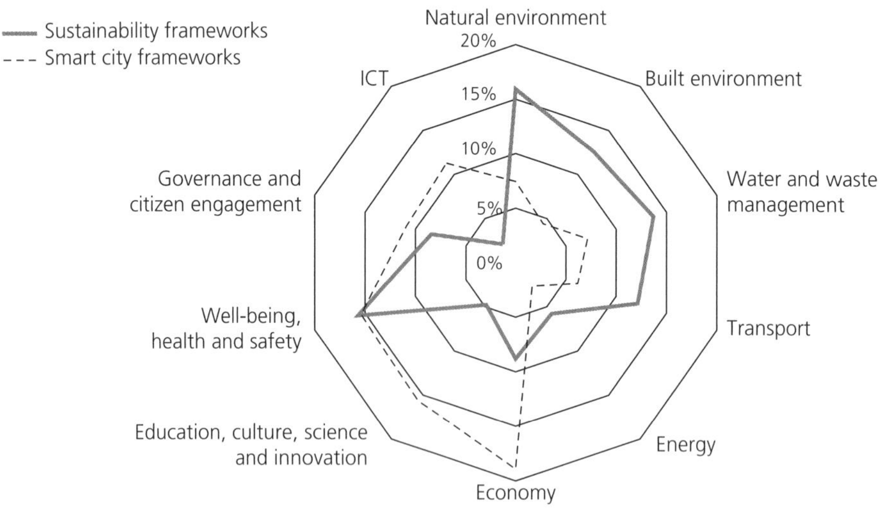

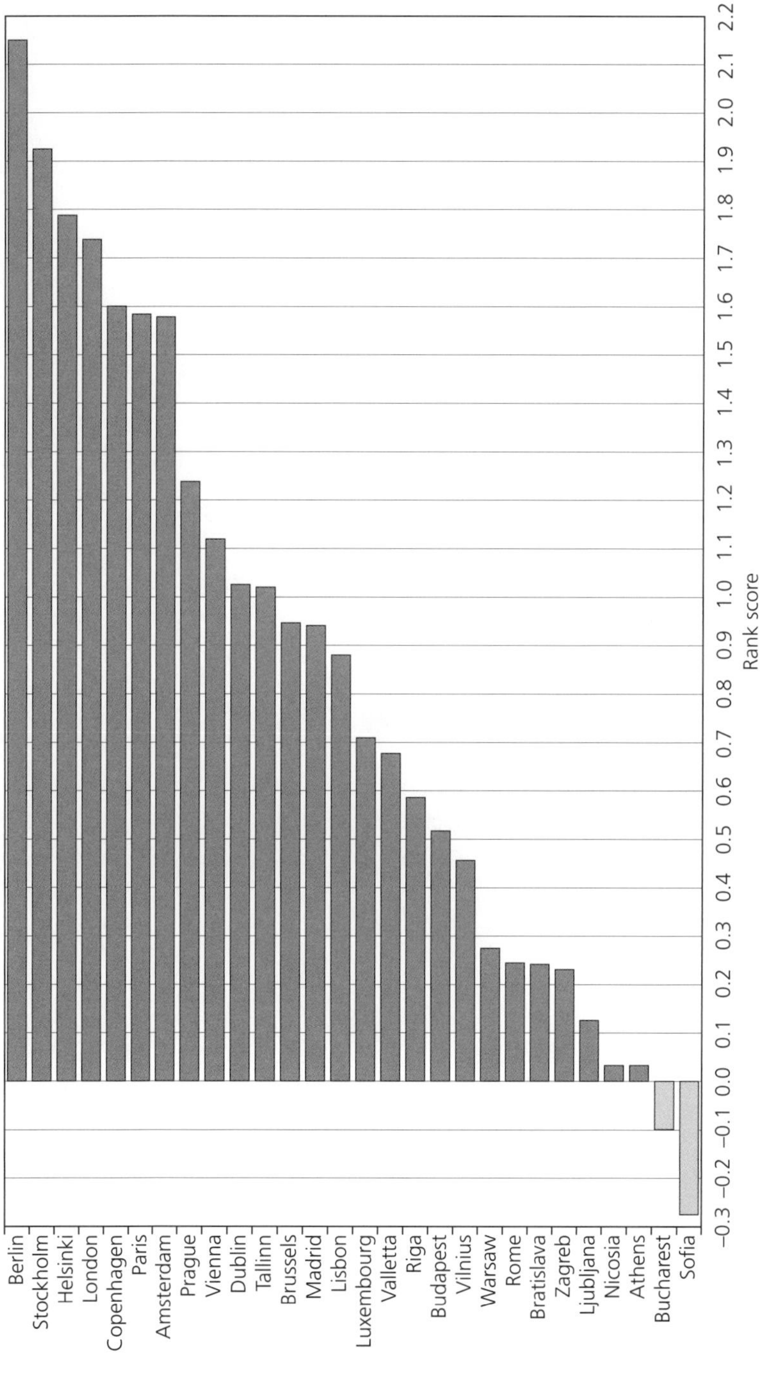

Figure 3.5 The smart sustainable rank scores of European capital cities. (Source: Akande *et al.*, 2019, licensed under CC BY)

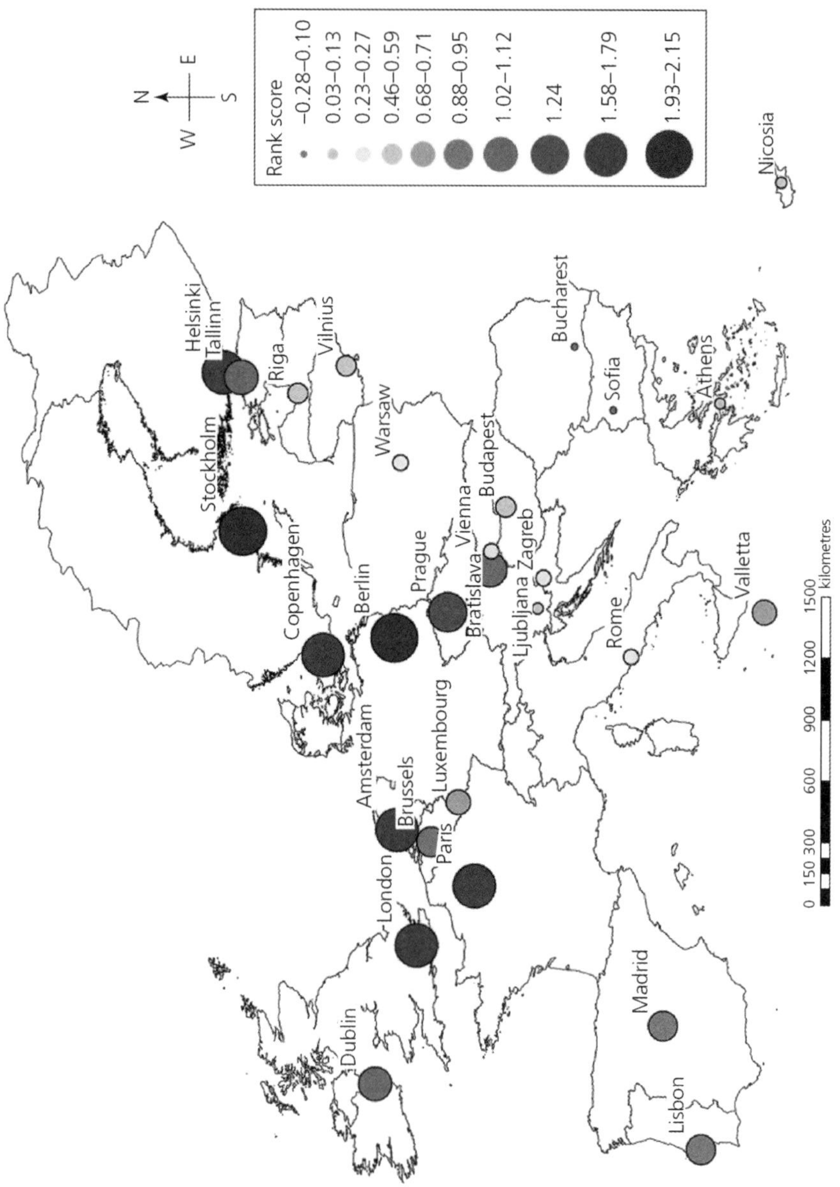

Figure 3.6 The smart sustainable ranking of European capital cities. (Source: Akande *et al.*, 2019, licensed under CC BY)

Akande *et al.* (2019) include indicators of sustainable energy and mobility, for example

- length of the bicycle network (dedicated cycle paths and lanes) (km)
- number of days that particulate matter PM_{10} concentrations exceed 50 $\mu g/m^3$
- greenhouse gas emissions from transport (million tonnes)
- share of the urban waste water load (in population equivalents) treated according to the applicable standard (%)
- proportion of the population living in households considering that they suffer from noise (%)
- electricity generated from renewable sources (%).

However, examining their analysis of the breakdown of those indictors contributing to the overall index score for each city shows the contribution of measures related to waste water treatment and air pollution measures to the score of each city, while other measures of activity related to the use of renewables and greenhouse gas emissions had low impact on the overall city index scores. Akande *et al.* (2019) noted that those measures that impact on a city's overall index score can serve as leverage points for city councils and urban planners to focus on. Their work highlighted that smart energy measures were not key 'leverage' measures for any of the European cities mentioned.

This section has highlighted both the diversity of activity within smart cities and how different cities may prioritise particular aspects of smart city activity, according to available local resources, expertise and legacy. Smart cities may not align closely with the stylised ideals of innovation clusters, or innovation systems, although some characteristics are clearly shared. The next section explores the activities of environmental firms located in smart cities, to ask whether we can observe any distinctive forms of activity.

3.4.1 Are environmental firms located in smart cities different?

Smart city initiatives have progressed at different rates across countries, and within them. Akande *et al.* (2019) have suggested a pattern of more accelerated development in northern European countries. In this section we focus on activity within a single country, the UK. Within the UK, London and Bristol have been identified as leading examples of smart cities (Woods *et al.*, 2016), with other challenger cities increasingly adopting smart city strategies and policy, with an emphasis on innovation and firm creation (e.g. see Newcastle Helix, 2019). Public sector support, from initially the Technology Strategy Board, and its successor Innovate UK, have been targeted at stimulating innovation to solve city challenges, through funding competitions open to UK cities. However, Buck and While (2017) reported that such projects have often relied on pre-existing public–private relationships to deliver existing solutions, rather than develop innovative projects and engage broadly with the private sector operating in city locations, such as those proposed in city lab initiatives. Observations from Buck and While (2017) challenge the idea that smart cities can be sources of innovative and entrepreneurial activity, instead positioning them as locations for the adoption of existing technology from a limited number of international corporations. However, as Cowley and Caprotti (2018) found in recent case studies, smart city activity in the UK is emergent and experimental, rather than standardised via top-down policy. Building on the observations of Shelton *et al.* (2014), we also noted that smart cities are inevitably the output of existing 'legacy' city characteristics, including social, built environment and governance arrangements. We investigated these ideas by examining whether the location of smart environment policy had a relationship to the underlying pre-existing environmental entrepreneurship base in the UK.

Using a firm level dataset of 330 SMEs with less than 250 employees from the UK's environmental sector, we undertook an exploratory investigation of locations for environmental entrepreneurship in the UK. Data were collected in 2010–2011 via a structured survey, investigating the characteristics of innovative and entrepreneurial activity. From this sample we matched 88 firms in 21 UK city locations with activity corresponding to the smart environment from the Manville *et al.* (2014) analysis of European smart cities. The remaining 242 firms were distributed across UK locations (city and towns), not listed at the time as having smart city environment policies, including city locations such as Bath, Cardiff and Leeds, as well as UK town locations. For Manville *et al.* (2014), 'smart environment' included smart energy, renewables, ICT-enabled energy grids, metering, pollution control and monitoring, renovation of buildings and amenities, green buildings and green urban planning, as well as resource use efficiency, re-use and resource substitution. It also included urban services such as street lighting, waste management, drainage systems, and water resource systems that are monitored to evaluate the system, reduce pollution and improve water quality.

Using descriptive statistics and simple *t* tests (the Welch and Student's test gave similar results) we compared those 88 firms based in cities with smart environment initiatives, with the remainder of firms based in non-smart city locations across the UK. Figure 3.7 presents the mean scores of both groups for different innovation and entrepreneurial characteristics. Firms were asked to report on the importance of different activities, using a five-point Likert scale. As smart city initiatives tend to target large urban areas, such as London, Manchester and Glasgow, one cautionary point is that differences in the two sample means may reflect underlying differences in the size of locations where firms are situated.

Responding firms in smart cities were on average more likely to be involved in collaborative activity, viewing collaboration as more important than firms based outside smart city locations. In particular, firms in smart environment city locations had a higher level of collaboration with universities and research centres for innovative activity. This relates to the observation that they were more likely to purchase R&D-related services as part of their core business activity, focus on in-house product innovation as part of their core business strategy and give a higher rating to applied R&D compared with other firms. For other characteristics examined, including the type of funding arrangement, demand for finance, barriers to funding, innovative outputs and use of skilled labour, we found no difference between firms based in smart city locations. However, firms based in smart cities had a self-perception of being more entrepreneurial and requiring venture capital more than those located outside smart cities.

Figure 3.7 Characteristics of firms grouped according to location (smart city compared with non-smart city)

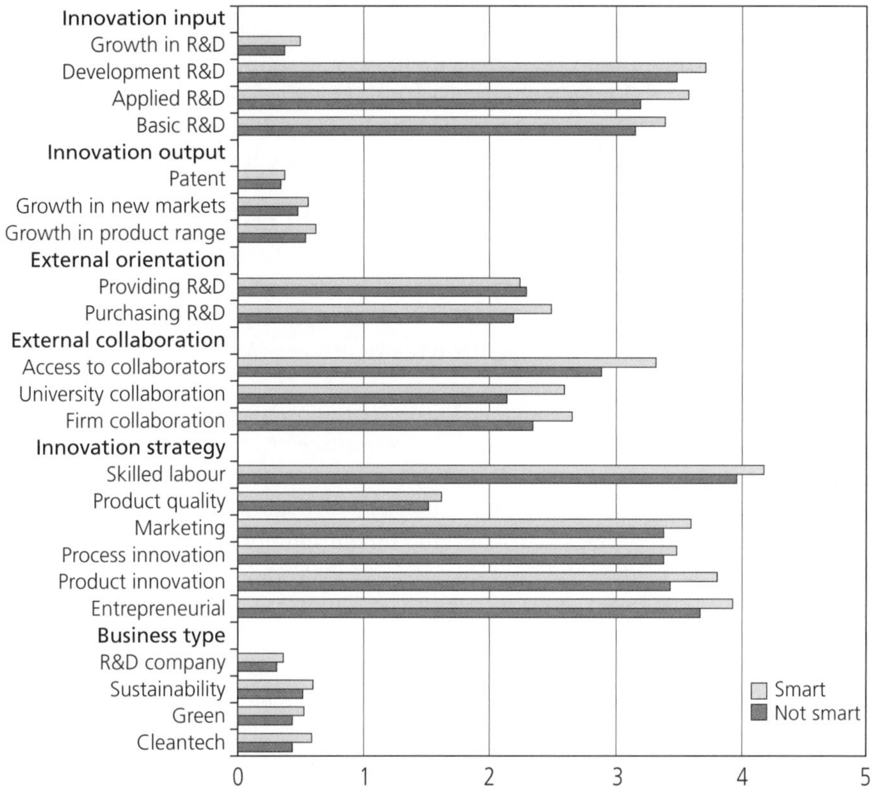

Several recent surveys of European research managers have highlighted the challenge of encouraging collaboration between academia and industry (Maukola, 2015). These results indicate the possibility of smart city initiatives to support such activities. The greater engagement of firms in smart cities actively purchasing R&D services may also contribute to the social and relational capital necessary to support innovative activity, particularly if this is via collaboration. This is also perhaps in contrast to our finding that providing R&D services is viewed as of 'relatively low importance' for all firms in the sample, since a service-based R&D strategy has been noted as a useful way for firms to reduce the uncertainty of early-stage R&D while simultaneously overcoming the challenge of searching for early-stage financing (Probert *et al.*, 2013).

These results are exploratory and indicate that smart city locations may be associated with a distinctive pattern of innovative firm characteristics, and indicate the potential for firms to engage in an innovation system via collaboration and to develop the critical mass of social and relational capital needed to stimulate further innovation. Our analysis in Table 3.1 lacked a measure of the social capital–social network activity, and this is likely a limitation.

Similarly to the results in Table 3.1, the results in Figure 3.7 are likely to reflect the strategy of selecting smart city locations and initiatives in alignment with pre-existing activity, rather than the direct influence of smart city policy. One limitation of this analysis is the absence of any longitudinal survey data to explore long-term dynamics. Several smart cities in the analysis, such as London and Cambridge, have a long history of innovative activity, and targeted policy support. However, observing that smart city initiatives focus on places with an established innovative and entrepreneurial trajectory also recognises the path-dependent nature of the development of innovative locations.

3.5. Network approach to environmental entrepreneurship

The discussion of the value of networks and social capital shares a similar perspective to 'network approach to entrepreneurship' proposed by Brüderl and Preisendörfer (1998), and is likely important to environmental entrepreneurship, particularly in smart city locations where engaging the local population has also been key.

Researchers in the field of entrepreneurship have long noted that networks make an important contribution to entrepreneurial performance. For example, entrepreneurs with larger, more diverse networks will be able to obtain more support for their projects and ventures than those entrepreneurs with small networks (Witt, 2004). Sorenson (2003) sees social relationships as important for an entrepreneur to access a range resources such as tacit industry knowledge and human and financial capital. In the network success hypothesis, networks based on founders' personal and business relationships can be used to acquire information and resources that may generally be unavailable or more expensive if obtained in a market (Brüderl and Preisendörfer, 1998). In this sense, the distinction between social capital and personal capital is clear (Burt, 1992). Personal or human capital rests with the individual's possessions, skills abilities and money, while social capital is based on relationships (Lin, 2002). Social relationships come 'into existence when individuals attempt to make best use of their individual resources' (Coleman, 1990, p. 300). Although social capital resides in relationships, it can be seen as providing access to resources that can be used by the individual.

From an entrepreneurship perspective, social capital can be leveraged by the entrepreneur to improve their access to resources and therefore enhance the performance of their business. Social capital can influence the success of actions for an individual; for example, individuals investing in social capital may benefit from the flow of information from other agents. This can open up new opportunities or help reduce potential search and transaction costs. By creating ties with others, an individual increases their access to information from new sources or overcomes imperfections in the market in terms of geographies or hierarchies (Burt, 1992; Granovetter, 1973, 1985).

Social ties also have the potential to influence critical decision-making. Social ties to the right individuals may carry weight in terms of credibility. A tie to an influential individual may be important if they are prepared to 'put in a word' on your behalf (Lin, 2002). As an example, an entrepreneur with a relationship to an important industry contact may benefit directly from their knowledge and experience, but may also gain access to other relevant associates or resources as a result of the social tie. Burt (1992) attributed value to connections that create competitive advantages by providing information and access to resources that would be unlikely to become available otherwise. They can be particularly valuable if these relationships are generally unavailable to others, and therefore overcome 'structural holes' where few relationships exist in a network.

Diverse networks can support the entrepreneur in the production and delivery of goods and services. In the initial phases of development, a start-up founder is likely to use their current social capital and a trial-and-error approach to develop their network of ties on an interpersonal level (Larson and Starr, 1993). The connections that prove useful to the entrepreneur in running their business become of value and reflect the entrepreneur's social capital. In this way, relationships become a useful resource for the firm (Larson and Starr, 1993). Entrepreneurs can adapt and manipulate their network in terms of size and access to resources and information by building the right relationships to suit their needs (Greve and Salaff, 2003). In particular, having contact with other entrepreneurs operating at a national or international level can provide access to a range of contacts and improve business growth (Donckels and Lambrecht, 1995).

Cohen and Fields (2000) see social capital as facilitated by trust, where trust is based on performance, reliability and reputation; for example, they view social capital as, 'focused, productive interactions among ... social institutions, instruments and entities' (Cohen and Fields, 2000, p. 193), which include universities, venture capitalists, law firms and business networks influenced by the structure of local industry. In their view, social capital is maintained through the productivity or performance of ties. It follows that a natural part of sustaining relationships and the utility of any social capital represented by them is a certain amount of obligation between tied actors, in terms of reciprocity or compensation (Coleman, 1990).

The utility of social networks is thought to be one explanation for the ability of entrepreneurial firms to grow and to compete with larger firms. The flexibility of informal relationships means that entrepreneurs can hold on to useful connections and build long-term relationships (Johannisson, 1996, citing Schutjens and Stam, 2001). The ability to develop social capital to provide access to resources can compensate for the liabilities of the start-up in terms of their size, such as restricted firm-level resources and a lack of historical legitimacy. Applying the network success hypothesis, those that access resources more cheaply than their competitors could via the market will be more successful (Witt, 2004).

Our analysis in Table 3.1 gave some support to the role of universities and individual entrepreneurs in supporting smart environment activities in cities. The triple-helix concept – linking public and private actors with universities as central actors – seems relevant to the smart cities studied here. Indeed, Grimaldi (2016) suggested that the smart city represents a new form of entrepreneurship, characterised by a quadruple helix of public administrators, private firms, universities and citizen engagement. Here, this represents a substantial shift in the perspective of the social capital of entrepreneurship, as previous research has assumed social capital is leveraged by entrepreneurs for their own, often commercial, purpose. In contrast, the inclusion of citizens within the network creates a different dynamic, engaging with those impacted by smart city activities, and so have their own set of interests. It suggests that social capital in smart cities should be considered a collective asset. In this sense it moves us closer to Robert Putnam's definition of 'civic virtue', which if embedded with communities can serve as a powerful tool for community development (Putnam *et al.*, 1993).

However, there are contrasting views of social capital as an individual, group, or collective asset that leads to different views on its value (Portes, 2000). Placing too much emphasis on social capital without understanding the underlying mechanisms risks misunderstanding the factors that drive the formation and development of such networks in the first place, or which may not be consistent from city to city.

3.6. Conclusion

Previous research has demonstrated the strong potential for smart cities to be a location for innovation and entrepreneurship. We have found mixed evidence on the link between smart cities and environmental entrepreneurship, perhaps reflecting the diversity of cities adopting such policies. From our exploratory research, we have seen that even in regions with concentrations of resources associated with providing supportive innovative environments, these alone are not sufficient explanations for the location or characteristics of smart cities, although there will be clear advantages for entrepreneurs based in such locations. We found some links between entrepreneurial activity and smart environment policy, but the relationship between smart cities and the presence of an underlying innovation system appears weaker, lacking connection to the key characteristics of smart city activities, such as high-technology activity, ICT and human capital. This may reflect both the diversity of smart cities and the breadth of smart city policy, which include a range of activity of which enterprise is a subset. It is clear that smart cities cover a wide range of city characteristics, with individual cities having specific legacies, capabilities and challenges that shape the reasons and needs for adopting smart city policy. As an example, a legacy of manufacturing activity appears to be linked with early adopters (mature smart cities) and the pursuit of smart environment policy.

In contrast, focusing on the UK, we have found some evidence of firms in smart city locations with greater engagement in innovative activity and being more collaborative. It is probable that these locations were already innovative places before becoming smart cities, but the presence of network activity in such places, identified as vital for innovation and smart cities, will likely support cities to achieve superior innovative and entrepreneurial performance. However, existing research still points to a challenge for UK smart cities to develop their own innovative solutions, and this potentially risks smart cities being adopters of 'imported' technology and approaches arriving from elsewhere. Again this may challenge the functioning of smart cities to achieve the 'quadruple helix', including engaging with local issues and people, and building solutions from the ground up.

The engagement of citizens and network activity within smart cities deserves further attention. In smart cities, the engagement of citizens is important, as it is these actors whose day-to-day livelihoods, surroundings and wellbeing are to be improved by smart city policy. The involvement of citizens within the innovative process and the development of networks will be key to achieving smart city aspirations. Smart city definitions place ICT and network communications at their core, but a key challenge is for cities to develop a tradition of 'civic spirit' built around social networks to encourage the exchange of knowledge and the development of trust to support smart city development. There is a clear role for multi-level policy to support the development of smart city locations, targeting the high-level infrastructure and resources, but also building networks to bring together individuals from the public sector and the private sector and local citizens who may otherwise remain unconnected in order to leverage their full potential.

Part 4 of this book will focus on governance issues relating to smart cities. The next chapter considers the Swedish experience of sustainable development and smart communities.

REFERENCES

Ahvenniemi H, Huovila A, Pinto-Seppä I and Airaksinen M (2017) What are the differences between sustainable and smart cities? *Cities* **60**: 234–245.

Akande A, Cabral P, Gomes P and Casteleyn S (2019) The Lisbon ranking for smart sustainable cities in Europe. *Sustainable Cities and Society* **44**: 475–487.

Berry CR and Glaeser EL (2005) The divergence of human capital levels across cities. *Papers in Regional Science* **84(3)**: 407–444.

Bosetti V, Carraro C, Galeotti M, Massetti E and Tavoni M (2009) Optimal energy investment and R&D strategies to stabilize atmospheric greenhouse gas concentrations. *Resource and Energy Economics* **31(2)**: 123–137.

Bower D (2003) Business model fashion and the academic spin out. *R&D Management* **33(3)**: 97–106.

Brüderl P and Preisendörfer P (1998) Network support and the success of newly founded businesses. *Small Business Economics* **10**: 213–225.

Buck NT and While A (2017) Competitive urbanism and the limits to smart city innovation: The UK Future Cities initiative. *Urban Studies* **54(2)**: 501–519.

Bürer MJ and Wüstenhagen R (2009) Which renewable energy policy is a venture capitalist's best friend? Empirical evidence from a survey of international cleantech investors. *Energy Policy* **37(12)**: 4997–5006.

Burt RS (1992) *Structural Holes: The Social Structure of Competition.* Harvard University Press, London, UK.

Cantwell J and Iammarino S (2000) Multinational corporations and the location of technological innovation in the UK regions. *Regional Studies,* **34(4)**: 317–322.

Caragliu A, Del Bo C and Nijkamp P (2011) Smart cities in Europe. *Journal of Urban Technology* **18(2)**: 65–82.

Centre of Regional Science, Vienna University (2007) *Smart Cities: Ranking of European Medium-sized Cities.* Centre of Regional Science, Vienna University, Vienna, Austria. http://www.smart-cities.eu/download/smart_cities_final_report.pdf (accessed 19/06/2019).

Cohen SS and Fields G (2000) Social capital and capital gains: an examination of social capital in Silicon Valley. In *Understanding Silicon Valley* (Kenney M (ed.)). Stanford University Press, Palo Alto, CA, USA.

Coleman JS (1990) *Foundations of Social Theory*. Harvard University Press, Cambridge, MA, USA.

Cooke P (2001) Regional innovation systems, clusters, and the knowledge economy. *Industrial and Corporate Change* **10(4)**: 945–974.

Cooke P (2004) Regional knowledge capabilities, embeddedness of firms and industrial organisations: bioscience Megacentres and economic geography. *European Planning Studies* **12(5)**: 625–641.

Cooke P (ed.) (2007) Regional knowledge economies: markets, clusters and innovation. Edward Elgar, Cheltenham, UK.

Cowley R and Caprotti F (2018) Smart city as anti-planning in the UK. *Environment and Planning D: Society and Space* **37(3)**: 428–448.

Cullen SE (2009) *Alternative Energy Powers Up: Staking Out the Patent Landscape for Energy from Wind, Sun and Waves*. Thomson-Reuters, Toronto, Canada. http://lib.sioc.ac.cn/tsg_admin/upload/myupload_4514.pdf (accessed 19/06/2019).

De Jager D, Klessmann C, Stricker E *et al.* (2011) *Financing Renewable Energy in the European Energy Market*. Ecofys, London, UK. https://ec.europa.eu/energy/sites/ener/files/documents/2011_financing_renewable.pdf (accessed 20/06/2019).

de Jong M, Joss S, Schraven D, Zhan C and Weijnen M (2015) Sustainable–smart–resilient–low carbon–eco–knowledge cities; making sense of a multitude of concepts promoting sustainable urbanization. *Journal of Cleaner Production* **109**: 25–38.

Donckels R and Lambrecht J (1995) Networks and small business growth: an explanatory model. *Small Business Economics* **7**: 273–289.

European Commission (2011) Helping the green sector through difficult times, Environment: Eco-innovation Action Plan. https://ec.europa.eu/environment/ecoap/about-eco-innovation/policies-matters/eu/386_en (accessed 20/06/2019).

Etzkowitz H and Leydesdorff L (2000) The dynamics of innovation: from national systems and 'mode 2' to a triple helix of university–industry–government relations. *Research Policy* **29(2)**: 109–123.

Eurostat (2012) Science and technology database. http://epp.eurostat.ec.europa.eu/portal/page/portal/science_technology_innovation/data/database (accessed 2012).

Eventure Media (2009) *The Development of Clean Technologies in the UK*. Eventure Media, London, UK.

Feldman MP and Francis JL (2003) Fortune favours the prepared region: the case of entrepreneurship and the Capitol region biotechnology cluster. *European Planning Studies* **11(7)**: 765–788.

Florida R (2002) The economic geography of talent. *Annals of the Association of American Geographers* **92(4)**: 743–755.

Foxon TJ, Gross R, Chase A, Howes J, Arnall A and Anderson D (2005) UK innovation systems for new and renewable energy technologies: drivers, barriers and systems failures. *Energy Policy* **33(16)**: 2123–2137.

Freeman C (1991) Networks of innovators: a synthesis of research issues. *Research Policy* **20(5)**: 499–514.

Gargiulo Morelli V, Weijnen M, Van Bueren E *et al.* (2013) Towards intelligently-sustainable cities? *TeMA – Journal of Land Use, Mobility and Environment* **1**: 73–86.

Garnsey E, Stam E and Heffernan P (2006) New firm growth: exploring processes and paths. *Industry and Innovation* **13(1)**: 1–20.

Granovetter MS (1973) The strength of weak ties. *American Journal of Sociology* **78(6)**: 1360–1380.

Granovetter MS (1985) Economic action and social structure: the problem embeddedness. *American Journal of Sociology* **91(3)**: 481- 510.

Greve A and Salaff JW (2003) Social networks and entrepreneurship. *Entrepreneurship Theory and Practice* **28(1)**: 1–22.

Grimaldi D (2016) *New Forms of Entrepreneurship and Innovation for Developing Smart Cities*. PhD thesis, Polytechnic University of Catalonia (UPC), Barcelona, Spain. https://upcommons.upc.edu/bitstream/handle/2117/105833/TDG1de1.pdf?sequence = 1&isAllowed = y (accessed 19/06/2019).

Hernández Guevara H, Tübke A and Brandsma A (2008) *The 2008 EU Industrial R&D Investment Scoreboard*. DG Research – Joint Research Centre, Institute for Prospective Technological Studies, Seville, Spain.

Hollands RG (2008) Will the real smart city please stand up? Intelligent, progressive or entrepreneurial? *City* **12(3)**: 303–320.

Iammarino S and McCann P (2006) The structure and evolution of industrial clusters: Transactions, technology and knowledge spillovers. *Research Policy* **35(7)**: 1018–1036.

Jacobsson S and Johnson A (2000) The diffusion of renewable energy technology: an analytical framework and key issues for research. *Energy Policy* **28(9)**: 625–640.

Jacquier-Roux V and Bourgeois B (2002) New networks of technological creation in energy industries: reassessment of the roles of equipment suppliers and operators. *Technology Analysis and Strategic Management* **14(4)**: 399–417.

Jamasb T, Nuttal WJ, Pollitt MG and Maratou A (2008) Technologies for a low-carbon electricity system: an assessment of the UK's issues and options. In *Delivering a Low-carbon Electricity System: Technologies, Economics and Policy* (M Grubb, T Jamasb and MG Pollitt (eds)). Cambridge University Press, Cambridge, UK.

Johannisson B (1996) The dynamics of entrepreneurial networks. *Frontiers of Entrepreneurship Research 1999: Proceedings of the 19th Annual Entrepreneurship Research Conference* (Reynolds P, Bygrave WD, Manigart S *et al.* (eds)). Babson College, Wellesley, MA, USA, pp. 253–267

Keeble D, Lawson C, Moore B and Wilkinson F (1999) Collective learning processes, networking and 'institutional thickness' in the Cambridge region. *Regional Studies* **33(4)**: 319–332.

Kourtit K, Nijkamp P and Arribas D (2012) Smart cities in perspective – a comparative European study by means of self-organizing maps. *Innovation: The European Journal of Social Science Research* **25(2)**: 229–246.

Kummitha RKR (2018) Entrepreneurial urbanism and technological panacea: why smart city planning needs to go beyond corporate visioning. *Technological Forecasting and Social Change* **137**: 330–339. https://doi.org/10.1016/J.TECHFORE.2018.07.010 (accessed 19/06/2019).

Larson A and Starr JA (1993) A network model of organizational formation. *Entrepreneurship Theory and Practice* **17**: 5–15

Lazaroiu GC and Roscia M (2012) Definition methodology for the smart cities model. *Energy* **47(1)**: 326–332. https://doi.org/10.1016/J.ENERGY.2012.09.028 (accessed 19/06/2019).

Lee B, Iliev I and Preston F (2009) *Who Owns Our Low Carbon Future? Intellectual Property and Energy Technologies*. Chatham House, London, UK. http://www.chathamhouse.org.uk/publications/papers/view/-/id/775 (accessed 19/06/2019).

Leibovitz J (2004) 'Embryonic' knowledge-based clusters and cities: the case of biotechnology in Scotland. *Urban Studies* **41(5/6)**: 1133–1155.

Lenox M and York JG (2011) Environmental entrepreneurship. In The Oxford Handbook of Business and the Natural Environment. (Pratima B and Hoffman AJ (eds)). Oxford University Press, Oxford, UK, 10.1093/oxfordhb/9780199584451.003.0004.

Lin N (2002) *Social Capital: A Theory of Social Structure and Action*. Cambridge University Press, Cambridge, UK, vol. 19.

Lombardi P, Giordano S, Farouh H and Yousef W (2012) Modelling the smart city performance. *The European Journal of Social Science Research* **25(2)**: 137–149.

Lundvall B (2000) Nation states and economic development-from national systems of Production to national systems of knowledge creating and learning. In *Oxford Handbook of Economic Geography* (Clark GL, Feldman MP and Gertler MS (eds)). Oxford University Press, Oxford, UK, pp. 353–372.

Lundvall B and Borrás S (1997) *The Globalising Learning Economy: Implications for Innovation Policy*. European Commission, Brussels, Belgium.

Madrid-Guijarro A, Garcia D and Van Auken H (2009) Barriers to innovation among Spanish manufacturing SMEs. *Journal of Small Business Management* **47(4)**: 465–488.

Manville C, Cochrane G, Cave J *et al.* (2014) *Mapping Smart Cities in the EU*. Directorate General for Internal Policies, European Parliament, Brussels, Belgium. http://www.europarl.europa.eu/RegData/etudes/etudes/join/2014/507480/IPOL-ITRE_ET(2014)507480_EN.pdf (accessed 19/06/2019).

Margolis RM and Kammen DM (1999) Evidence of under-investment in energy R&D in the United States and the impact of federal policy. *Energy Policy* **27(10)**: 575–584.

Marshall A (2009) *Principles of Economics*, 8th edition. Cosimo, New York, NY, USA.

Mason CM and Harrison RT (2004) Improving access to early stage venture capital in regional economies: a new approach to investment readiness. *Local Economy* **19(2)**: 159–173.

Maukola J (2015) Universities still shun industry collaboration. *Research Professional*. https://www.researchprofessional.com/0/rr/news/europe/horizon-2020/2015/3/Universities-still-shun-industry-collaboration.html (accessed 19/06/2019).

Meadows DH, Goldsmith EI and Meadows D (1972) *The Limits to Growth: Club of Rome*. Universe, New York, NY, USA.

Morosini P (2004) Industrial clusters, knowledge integration and performance. *World Development* **32(2)**: 305–326.

Mytelka LK (2004) Clustering, long distance partnerships and the SME: a study of the French biotechnology sector. *Innovation: The International Journal of Technology Management* **27(8)**: 791–808.

Nelson RR and Winter SG (1982) *An Evolutionary Theory of Economic Change*. Harvard University Press, Cambridge, MA, USA.

Newcastle Helix (2019) https://www.ncl.ac.uk/helix/about/ (accessed 19/06/2019).

Niosi J and Banik M (2005) The evolution and performance of biotechnology regional systems of innovation. *Cambridge Journal of Econometrics* **29**: 343–357.

Porter M (1990) *The Competitive Advantage of Nations*. Palgrave Macmillan London, UK.

Porter M (2000) 'Location, competition, and economic development: local clusters in a global economy. *Economic Development Quarterly* **14(1)**: 15–34

Portes A (2000) The two meanings of social capital. *Sociological Forum* **15(1)**: 1–12.

Probert J, Connell D and Mina A (2013) R&D service firms: the hidden engine of the high-tech economy? *Research Policy* **42**: 1274–1285.

Putnam RD, Leonardi R and Nanetti R (1993) *Making Democracy Work?: Civic Traditions in Modern Italy*. Princeton University Press, Princeton, NJ, USA.

Richter C, Kraus S and Syrjä P (2015) The Smart City as an opportunity for entrepreneurship. *International Journal of Entrepreneurial Venturing* **7(3)**: 211.

Rosenberg N and Nathan R (1982) Inside the Black Box: Technology and Economics. Cambridge University Press, Cambridge, UK.

Senker J (1995) Tacit knowledge and models of innovation. *Industrial and Corporate Change* **4(2)**: 425–447.

SETIS (Strategic Energy Technologies Information System) (2019a) European initiative on smart cities. https://setis.ec.europa.eu/set-plan-implementation/technology-roadmaps/european-initiative-smart-cities (accessed 19/06/2019).

SETIS (2019b) Smart cities indicative roadmap. https://setis.ec.europa.eu/sites/default/files/basic_page/Smart%20cities_roadmap_flipped.jpg (accessed 19/06/2019).

Shelton T, Zook M and Wiig A (2014) The 'actually existing smart city'. *Cambridge Journal of Regions, Economy and Society* **8(1)**: 13–25.

Shrivastava P (1995) Environmental technologies and competitive advantage. *Strategic Management Journal* **16**: 183–200

Siemens (2008) *Sustainable Urban Infrastructure. London Edition – A View to 2025*. Siemens, Munich, Germany. https://w5.siemens.com/web/si/sl/corporate/portal/raziskave/Documents/sustainable_urban_infrastructure-study_london.pdf (accessed 19/06/2019).

Sorenson O (2003) Social networks and industrial geography. *Journal of Evolutionary Economics* **13**: 513–527.

Storper M (1993) Regional 'worlds' of production: Learning and innovation in the technology districts of France, Italy and the USA. *Regional Studies* **27(5)**: 433–455.

Stuart TE and Ding WW (2006) When do scientists become entrepreneurs? The social structural antecedents of commercial activity in the academic life sciences. *American journal of Sociology* **112(1)**: 97–144.

Van Aalst HF (2003) *Networking in Society, Organisations and Education. Networks of Innovation*. Organisation for Economic Co-operation and Development, Paris, France.

Viitanen J and Kingston R (2014) Smart cities and green growth: outsourcing democratic and environmental resilience to the global technology sector. *Environment and Planning A: Economy and Space* **46(4)**: 803–819.

Von Burg U and Kenney M (2000) Venture capital and the birth of the local area networking industry. *Research Policy* **29**: 1135–1155.

Witt P (2004) Entrepreneurs' networks and the success of start-ups. *Entrepreneurship and Regional Development* **16(5)**: 391–412.

Woods E, Alexander D, Labastida RL and Watson R (2016) *UK Smart Cities Index. Assessment of Strategy and Execution of the UK's Leading Smart Cities*. Navigant Consulting, Boulder, CO, USA. https://www-file.huawei.com/-/media/corporate/pdf/news/huawei_smart_cities_report_final.pdf?la=en&source=corp_comm (accessed 19/06/2019).

ICE Publishing

Nuttall, William J, Gibson, David V, Trzmielak, Dariusz and Ibarra-Yunez, Alejandro
ISBN 978-0-7277-6425-6
https://doi.org/10.1680/emsc.64256.061

Chapter 4
Sustainable development and smart communities: the Swedish experience

Sheridan Tatsuno
OneReality.com, Stockholm, Sweden

4.1. The climate challenge: designing carbon-neutral cities under the Paris Accord

The United Nations (UN) International Panel on Climate Change recommends that we must establish the framework, policies and programmes to limit global warming to 1.5°C by 2030. How can we reach this goal? Most policy discussions have focused on international and national policies, but have overlooked cities, which play a major role since they consume 80% of the energy we produce and emit over 70% of the carbon: the climate battle will be won or lost in the cities. For example, in the USA, traffic congestion, which is largely concentrated in cities (McCarthy, 2019), contributes to 27% of the nation's carbon dioxide emissions and is projected to increase by 50% by 2030 (Torres, 2017). However, unlike national governments whose policies are heavily influenced by fossil fuel companies, cities are energy end-users. They may operate municipal energy utilities and are significant energy buyers – who determine planning, traffic, zoning and energy-purchasing policies – and they guide all construction projects, which will rise to $12.9 trillion in 2022, up from $10.8 trillion in 2017, or currently about 11% of the global economy (Reportlinker, 2018). Thus, they have enormous power to influence renewable energy purchasing for public buildings and transit electrification. However, many cities encounter difficulties in encouraging private homeowners, especially senior citizens and low-income families, to install solar panels. Instead, cities like Palo Alto, California, are buying energy from solar community grids on a competitive basis, or, like San Francisco, establishing their own city-owned clean energy company. City-level policy, and the purchasing power that follows, can reduce energy use, traffic congestion and carbon emissions through smart city and mobility technologies, renewable energies and sustainable development (SD).

Currently, cities and technology vendors are promoting smart cities and smart mobility as ways to reduce costs, improve traffic flow and surveillance, and generate new usage revenues. While effective at minimising costs for petrol-powered vehicles, they are not truly 'smart' since they are increasing carbon emissions by generating more traffic. A truly 'smart' city would aim to reduce long-distance commuting and adopt renewable energies and electric vehicles (EVs), but this will be economically difficult, especially in oil-producing nations, as well as highly likely to be met with social resistance.

Moreover, climate change requires resilience planning to minimise disasters. The need for resilience planning makes shifting to renewables urgent but also costly. How can cities finance these costly investments while shifting to renewable energies?

Sweden, rated as the most sustainable nation in the world in 2015, provides policy insights from decades of implementing SD at the national and local community levels. Since the 1970s, Sweden has gradually shifted to low-carbon 'eco-cities' using a range of smart and sustainable technologies, showing that it is economically and socially viable to prepare cities, communities and individual citizens for life after fossil fuels (see Chapter 1). Indeed, the Harvard Business School forecasts that saving the planet from ecological disaster is a $12 trillion opportunity (Elkington, 2017).

How did Sweden achieve its SD goals? What SD initiatives did cities and towns undertake? Which policies worked and which did not? How did energy, mobility, and health and wellbeing fit within national SD policies and urban design? This chapter explores Sweden's experience in SD policies at a national and local level in terms of economics, environmental and social sustainability as proposed by the UN Millennium Declaration, with a particular focus on Sweden's emphasis on health, wellbeing, employment and quality of life. A 'circles of sustainability' approach is introduced, which includes the four domains of economic, ecological, political and cultural sustainability. The chapter explores emerging ideas about eco-villages, eco-towns and eco-cities, designed according to biophilic (the concept of people's inherent affinity for nature) principles.

4.2. Sweden's SD national strategy

Sweden has a long history of implementing climate policies through a variety of economic instruments such as carbon dioxide taxes, renewable energy certificates, subsidies, and research and development investments, legislation, carbon trading, international collaboration, voluntary agreements and dialogue between the state and businesses, and public participation. In the 1920s, Sweden imposed energy taxes on petrol and, during the high-growth 1950s, energy taxes on household, services and heating plants (85 €/MWh), tax exemptions on biomass fuels, and reduced taxes (25 €/MWh) on manufacturing, combined heat and power (CHP) plants, agriculture, forestry and aquaculture. The goal was to maximise energy efficiency, since Sweden relies on oil imports.

In 1967, Sweden became an early adopter of sustainability by establishing the first environmental protection agency, and implemented SD policies in cities and towns nationwide. In 1972, just prior to the OPEC oil embargo, Sweden hosted the first UN conference on the environment, which led to the creation of the UN Environment Programme – the leading global environmental authority to this day.

During the late 1970s, Sweden introduced industrial district heating using biofuels, peat, waste, natural gas and liquefied petroleum gas, and from the early 1980s waste heat, heat pumps and electric boilers. District heating is a major reason for Sweden's reduced greenhouse gas emissions in housing and the service sector, accounting for over 80% of heat and hot water used by apartment blocks. Sweden initially focused on environmental sustainability, but the post-Cold War recession forced the nation to reconsider its policies. According to Malin Ahlberg (2009)

> When the UN Conference on Environment and Development took place in Rio de Janeiro in 1992, Sweden's economy experienced its deepest crisis since the 1930s. In spite (or because) of the crisis the Swedish government initiated the political debate on sustainable development (SD) as part of far-reaching plans for structural change.

Rather than retreating, Sweden expanded its SD policies to include social, economic, environmental and other factors, with an emphasis on economic revitalisation and social justice. This policy shift was complemented by Swedish cities experimenting with innovative SD initiatives.

During the 1990s and early 2000s, Sweden broadened its energy and transport policies to more sectors (see Alopaeus, 2013), as listed below.

- Cross-sectoral: energy and carbon dioxide tax, EU Emissions Trading System, grants for local climate action programmes.
- Energy: energy and carbon dioxide tax (1991), investment grants, electricity certificates, voluntary agreements with industry, household energy efficiency. In 1991, a carbon dioxide tax was introduced, and companies began producing biofuels from peat, waste, bark, chips, flax and hemp and other industrial and farm wastes. In 2001, environmental tax increases on households and firms were balanced by tax cuts on labour. As a result of these policies, households and services were almost fossil fuel free in 2010.
- Transport: motor fuel taxes, 'green' cars for public procurement, carbon dioxide-differentiated vehicle tax, 5-year exemption from annual tax for green cars, investment grants for 'very green' cars (50 g of carbon dioxide per km). E85 fuelling stations were introduced, but in 2017, biofuels only accounted for 18.6% of fuels used in transportation (Svebio, 2017).
- Industrial processes: environmental codes, EU fluorinated gas regulation. Increased industrial energy use was balanced by decreasing emissions through bioenergy, energy/carbon dioxide taxes, electricity certificate schemes and higher fossil fuel prices.
- Waste management: landfill tax, landfill ban on combustible and organic waste.
- Emission trading (2005): emission allowance pricing depending on emission amounts.

Sweden was one of the first nations to sign and ratify the international climate change treaty and the Kyoto Protocol in 1998 and 2002, respectively. At The Stockholm Convention (2001), Sweden promoted a global treaty aimed at phasing out the production and use of persistent organic pollutants, as well as making progress with waste management, acid rain prevention, sustainable city planning and recycling (e.g. see Swedish Institute, 2019).

In 2003, SD was adopted as an overall policy objective after a 2002 survey showed that 95% of respondents favoured climate change tax spending in the aftermath of the 9/11 terrorist attacks. From 2003 to 2008, Sweden invested SEK 2 billion ($224 million) in climate grants to municipalities, primarily for biogas, district heating expansion, energy efficiency and transit programmes, which drove its climate and energy planning. However, the Swedish Environmental Protection Agency admitted that it would be difficult to evaluate the impact of this financial support, considering other factors such as its green taxes and higher fossil fuel prices. Nevertheless, Sweden expanded the Division for Sustainable Development at the Ministry of Environment and a 'Council for Sustainable Development' under the National Board of Building, Planning and Housing in 2005 to implement national SD policies at the local and regional levels (Ahlberg, 2009).

In 2006, in response to high oil prices due to the 9/11 attacks, Sweden created a Commission on Oil Independence with the ambitious 2020 targets of

- energy to be used 20% more efficiently, overall
- consumption of oil by industry to be cut by 25–40%
- consumption of oil by road transport to be reduced by 40–50%
- heating buildings with oil (a practice already cut by 70% since the 1973 oil crisis) to be phased out (e.g. see Atkisson, 2006).

In 2007, to balance SD with economic growth, the Commission on Sustainable Development was established. In 2008, Prime Minister Fredrik Reinfeldt recognised the need to balance climate policies with economic growth in his speech 'The new Swedish Model: a reform agenda for growth and the environment' (Reinfeldt, 2008)

> We believe that investments in research and new technology, together with fiscal and regulatory revisions, can pave the way for a development in which the environment becomes a lever for new enterprises and new jobs. A model for climate care and growth – this will be an important ingredient in the renewal of the Swedish Model).

The government introduced an annual sustainability report covering all 55 state-owned companies, which employed 190 000 people, as a model for sustainable growth. Sweden's national SD strategy was defined in holistic terms (cited in Ahlberg, 2009)

> The policy objectives of the vision of a sustainable society are solidarity and justice in every country among countries and among generations. The basic assumption is that members of one generation should not conduct their lives in a way that prevents their children or future generations from enjoying a decent standard of living. Sustainable development is an approach that must actively inform and shape all policy areas.

In 2009, Sweden identified four strategic challenges as priorities to realise its vision of a sustainable society, with jobs as central to all four challenges

- building sustainable communities and promoting qualify of life

- encouraging good health on equal terms for everyone
- meeting the demographic challenges (ageing, immigration, etc.)
- encouraging sustainable growth and recognising that growth is driven by dynamic markets, a forward-looking welfare policy, and a progressive environmental policy (Ahlberg, 2009).

Sweden adopted 87 social development indicators based on the work of Statistics Sweden, including 12 headline indicators (average life expectancy, violence, energy efficiency, investments, employment rate, public debt, growth, risk of poverty, demographic support ratio, greenhouse gases, hazardous substances and development assistance), which are broken down into six different areas (health, sustainable consumption and production, economic development, social cohesion, environment and climate, and global development). The indicators do not include all elements of SD, and the strategy does not include a rating system. Under the strategic challenge of building sustainable communities, Sweden pursued 16 national environmental quality objectives for long-term sustainability, which were adopted by the Swedish Parliament in 1999 until 2014 (Ministry of the Environment, 2006).

In 2011, Sweden promoted an environmental technology strategy to promote green-technology companies, which was backed by SEK 400 million (about $40 million) through 2014. For disaster preparedness, the Swedish Civil Contingencies Agency provides guidelines to help society adapt to climate change. Today, the sector employs 40 000 people and has revenues of SEK 120 billion ($12 billion), according to Statistics Sweden and the then Swedish Environmental Technology Council. The public has largely accepted these policies.

Swedish companies have actively developed corporate social responsibility (CSR) policies integrating social and environmental concerns, with 75% of the largest corporations involved. The leading CSR network platform in Sweden has 17 active members. The Confederation of Swedish Enterprise, which represents employer and trade associations, published the booklet *The Confederation of Swedish Enterprise View of Sustainable Development* in 2002.

Swedish universities are developing a variety of sustainability programmes and global partnerships for Swedish and foreign students. Chalmers University of Technology in Gothenburg is a partner in the Alliance for Global Sustainability with the Massachusetts Institute of Technology, the University of Tokyo and ETH Zürich. In Stockholm, KTH Royal Institute of Technology has invested SEK 8 billion (about 800 million USD) in the environment and SD through the KTH Sustainability Council.

In its 2012 roadmap to the year 2050, Sweden's Environmental Protection Agency set the goal of increasing renewable electricity to 25 TWh by 2025 out of a total electricity production of 150 TWh. The goal is to reduce transport carbon emissions by 70 TWh by 2030, then completely switch to fossil-free mobility. To increase vehicle fuel efficiency,

Sweden set the transport goals of

- an increased fuel tax and a strengthened carbon dioxide differentiated vehicle tax
- a European-wide voluntary agreement with car manufacturers on carbon dioxide emission from new cars
- joint EU regulation on carbon dioxide emission from new cars from 2012
- a 'very green' car bonus
- for biofuels: fuel tax exemption (1–2 SEK/kg of carbon dioxide), second-generation biofuels needed, filling stations obligated to provide alternative fuels, investment grants for biogas production and distribution, and lower vehicle taxes for flexible-fuel vehicles (Alopaeus, 2013).

Today, Sweden is the highest user of biofuels in Europe, at 32% of total fuel use, due to the commitment to E85, bioheating and bioelectricity for housing, services, industry and transport. In 2014, Sweden used 555 TWh of energy, 130 of which came from biofuels, reducing greenhouse gas emissions by 25% between 1990 and 2014 (e.g. see Sahlin, 2015).

As a result of its national SD policies, Sweden ranked first in the 2014 and 2016 editions of the Global Green Economy Index. The Scandinavian nation emerged ahead of 59 other countries evaluated on a broad range of environmental, social and governance components. According to the *OECD Environmental Performance Reviews 2014* (OECD, 2014), Sweden is one of the most innovative countries in environmental technologies, especially biofuels, smart grids and carbon capture and storage.

The environmental consultancy First Carbon Solutions believes there are five reasons why Sweden ranks at the top in terms of sustainability, as follows.

- **Sustainable living.** The Swedish government actively promotes sustainable living in a variety of areas, including organic foods and clothing, renewable energy and recycled beverage containers. Legislation supporting family-friendly policies such as maternity and paternity leave for employees was adopted.
- **Roadmap 2050.** Sweden intends to become the world's first fossil-fuel-free nation (Brittlebank, 2015) as part of its commitment under Roadmap 2050. Sweden aims to reduce greenhouse gas emissions by 40% from 1990 levels by the year 2020 and by 80% by 2050, and have a vehicle fleet completely rid of fossil fuels by 2030. To achieve 2020 goals, emissions need to decline by 16 million tonnes in 2020 through drastic measures.
- **Carbon tax.** In 1995, Sweden became the first nation to adopt a carbon tax. By 2012, tax revenues reached 2.52% of GDP, compared with the OECD average of 1.54%. By 2020, the government goal is to make energy use 20% more effective compared with 2008.
- **Renewable energy.** Sweden aims to boost renewable energy, largely hydropower and biofuels, from 52% today to 55% by 2020.
- **Human-heated buildings.** 'Passive houses' powered through energy from people's body heat, electrical appliances and sunlight are being built around the country (cited in Wilde, 2016).

Sweden also relied on six nuclear reactors in the 1970s, and on another six through 1985, with two units closed in 1999 and 2005. There are three operational nuclear power plants, with ten operational nuclear reactors producing 35–40% of the nation's electricity. Sweden pursued a phase-out policy, but in 2009 opted to retain the existing reactors. After the Fukushima meltdown in 2011, polls showed that 64% of Swedes opposed new reactors, while 27% supported them (Jonsson and Berglöf, 2016).

4.3. Sweden's sustainable cities

Within the context of Sweden's ambitious national sustainability policies, Swedish cities and towns have implemented SD in a variety of areas. There are 290 municipalities, 18 county councils and two regions – Västra Götaland and Skåne – that are self-governing with considerable autonomy and taxing powers (Figure 4.1). Expert commissions are appointed to study SD issues, with their recommendations open to opposing opinions. Swedish cities have implemented a variety of SD initiatives to improve their economies and quality of life.

4.3.1 Stockholm: climate policies and Hammarby eco-town

Stockholm is one of the most advanced sustainable cities in the world. Its climate efforts began with district heating in the late 1950s and biofuels during the mid-1960s. Its planning system and utilities infrastructure centred on an efficient public transportation system, green areas within a distance of less than 300 m from each residence, and dense city centre building, facilitating district heating and cooling, etc. Since 1995, two 5-year environmental plans (1995–2000 and 2000–2005) have been successfully implemented, with a new 10-year action plan launched in 2010. The city has long-term plans for becoming fossil fuel free by 2050. Its climate programme involves the entire city: Stockholm's climate initiatives are integrated in other processes for sustainable growth running in parallel, such as the Vision Stockholm 2030 project, which began in 2006 with a view to formulating a long-term complete vision for the development of the city, and the Walkable City, Stockholm's new city plan adopted in 2010. More than 125 companies are involved in the implementation of the city's plan through the Climate Pact, an initiative launched in September 2007 with a view to engaging business in local climate action. Among the key stakeholders of the initiatives are the citizens themselves: their involvement facilitates the implementation of measures beyond the City of Stockholm's direct mandate. Stockholm has implemented several communication projects facilitating the active involvement of citizens in the climate initiatives, such as the Climate Neutral Stockholmers campaign and the Climate Account (Global CCS Institute, 2013).

Located in Hammarby Lake City on Stockholm's South Island is the Smart City Sweden initiative, which is funded partly by the Swedish government through the Ministry of Enterprise and Innovation and the Swedish Energy Agency, and partly by industry and business-promoting networks. Managed by IVL Swedish Environmental Research Institute in close collaboration with business and other sustainability organisations, Smart City Sweden is a national export and investment platform for smart and sustainable city solutions. Hammarby is one of the most advanced eco-towns in Sweden (see Ignatieva and Berg, 2014).

Figure 4.1 Political map of Sweden

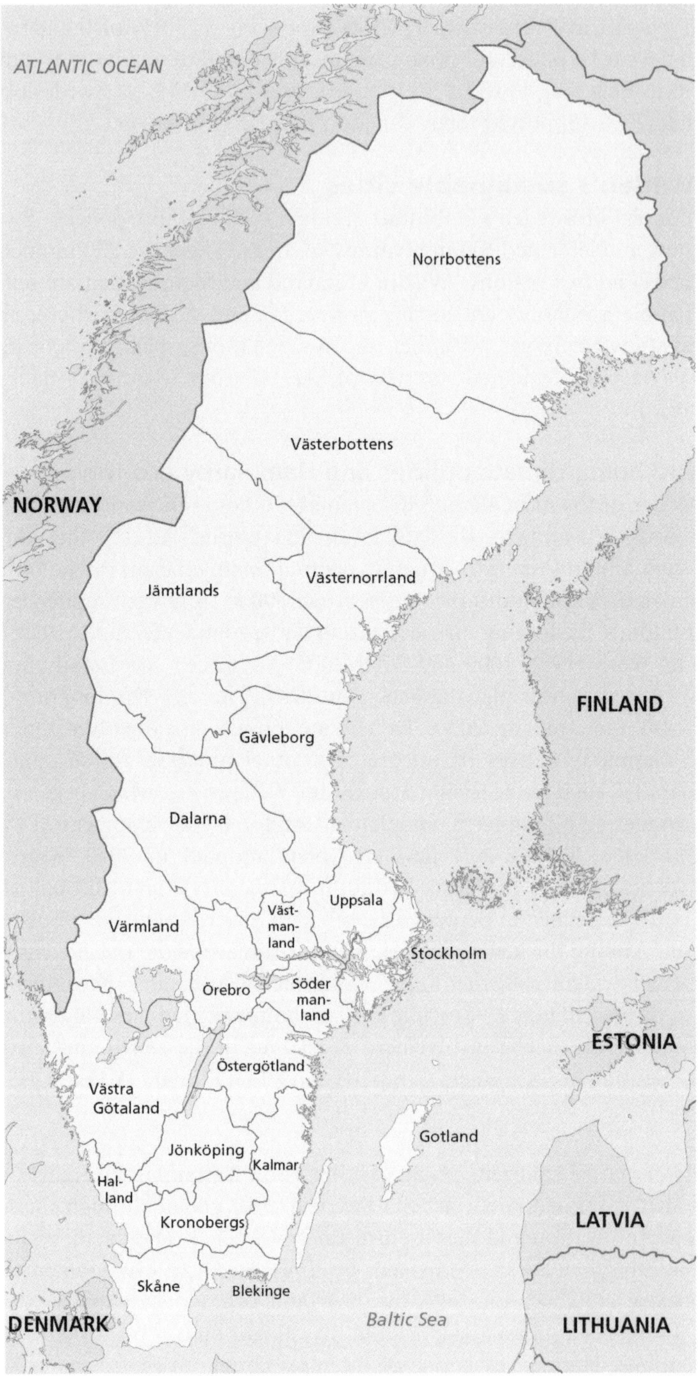

Originally intended to be a temporary Olympic Village in 2004, the former industrial area was redesigned as a bedroom eco-town for 35 000 workers commuting to Stockholm using public transport, bicycles and carpools. The Hammarby model reduced heat consumption by 50% and uses bioenergy and incinerated waste to co-generate heat and electricity. The four- and five-storey buildings have green roofs, solar panels, sunny green courtyards, small gardens and greenhouses, and are protected from winds. Hammarby's unique feature is its vacuum waste suction system, which sends household waste through underground pipes to substations to avoid garbage truck traffic. Food waste is converted to biogas, which, along with sludge from wastewater, fuels the town's buses. Local wastewater and storm water are harvested and filtered. Other features include a pedestrian and bicycle network, a large carpooling system, and a commuter ferry to downtown Stockholm (Ignatieva and Berg, 2014). Nearby is the Nacka Wedge, which has a large ski slope, vast forests, small fields and several lakes. Hammarby has become an eco-tourist destination, attracting over 13 000 visitors a year, and the Hammarby model is used by the Caofeiden Ecocity development in China and Symbiocity in Brazil. However, despite its popularity, Hammarby lacks many features that would make it a viable model for broader adoption: affordable rent, mid-size green areas, public squares for open markets and city life, cultural and income diversity, and a cultural centre. Its residents are mostly wealthier, non-immigrant families with children, which prevents it from becoming a model for social integration. Moreover, Hammarby was designed top down by planners and did not involve citizens in its planning, which is crucial according to biophilic principles.

4.3.2 Malmö's social green revitalisation

In 2013, the OECD ranked Malmö as the world's fourth most inventive city after Eindhoven, San Diego and San Francisco. It is one of Sweden's youngest and most international cities, with 183 nationalities. A small city of 330 000 people, of whom one-third are foreign born, Malmö has focused on developing green energy, digital healthcare, bicycling and sustainable tourism.

Located in southern Sweden across the Öresund Strait from Copenhagen, Malmö is a case study for urban planners interested in SD for economic revitalisation and growth (Fitzgerald, 2016). Like other industrial cities, Malmö suffered the collapse of its maritime and other heavy industries in the 1980s. Mayor Illmar Reepalu sought ways to revitalise the city, and learned that district heating, where all energy is recycled, could reduce energy waste; district heating recycles heat from industries and redistributes them to housing and other sectors. The city is a destination for cities learning how to implement total recycling using district heating and cooling.

During the 1990s, the city invested in emerging knowledge industries, opening business incubators and expanding its transportation networks. In 1995, Malmö University was founded to attract young people and lay the foundations for growth in information technology, education, economics, sustainability, design and media. In 2000, a bridge over the Öresund Strait to Denmark was built, tapping into Copenhagen's stronger labour market.

Malmö's Western Harbour is Europe's first carbon-neutral neighbourhood, with its heating and cooling run entirely on renewable energy. Waste is sent to a central underground tank using vacuum pipes. Food scraps are converted into biogas for buses. The Augustenborg Eco-City residential area, which has received the largest investments in Europe for ecological conversion, was awarded the UN World Habitat Award in 2010 for its 10 000 green roofs to absorb and recycle rainwater. Over 70% of its waste is recycled, and solar panels are installed on public and private buildings. The Hyllie smart city district will be 100% fuelled by renewable and recycled energy by 2020. Its sustainable energy system integrates electricity, heating and cooling, and residents can track and measure their environmental impact by using smart monitoring. Malmö is the world's sixth most bike-friendly city, with wide, dedicated paths connecting the entire city (Savage, 2017). Its vision is to become a world leader in sustainable city development by 2020 and shift to 100% renewable energy by 2030.

In the future, Malmö projects to add another 50 000 residents by 2030, while planning for continued SD and a high quality of life. The coastal district of Nyhamnen – which is close to green habitats and the central train/bus station and will have 6000 homes and 13 000 jobs when fully developed – is a model for future growth because it shows how cities can redesign abandoned industrial sites for new housing, office and retail uses. Malmö plans to spend SEK 4 billion ($400 million) on overall city planning, a network of electric buses and the new district of Pågatågstrafiken for about 30 000 new homes by the year 2035.

Unlike the 1960s, when growth focused on major gateways and large parking lots on the city outskirts, more recent developments focus on controlled growth and 'refinement', according to City Building Office and City Construction Director Christer Larsson. Slower growth is designed to avoid the problems caused when Sweden accepted larger numbers of Syrian refugees in 2015, causing a serious housing shortage, difficulties with inadequate social services and anti-immigrant sentiment in Malmö and other impacted towns and cities.

4.3.3 Gothenburg: No. 1 sustainable city

Gothenburg was ranked 'number 1' for sustainability and innovation in the 2017 Global Sustainability Index of world cities (Holmåker, 2017). Sweden's second-largest city (after Stockholm) is moving towards a low-carbon economy, which includes waste-fuelled heat and power plants, low-emission zones for heavy vehicles, cars fuelled by renewable biogas, and the world's first large-scale autonomous driving pilot project on public roads. Emission-free buses glide along Route 55 in ElectriCity – the sustainable transport demonstration project (see ElectriCity, 2019). In 2018, the city hosted the Vehicle Electronics and Connected Services event, to promote connected vehicles and automated driving.

Gothenburg's district heating network is 1200 km long, and heats 90% of the city's apartment blocks and 12 000 detached homes, further improving Sweden's reduced greenhouse gas emissions.

4.3.4 Lund's science village

With one of Europe's oldest universities, Lund is now becoming a centre of advanced scientific research with construction of the European Spallation Source (ESS) and MAX

IV particle accelerators. The ESS will generate 20% of the heat for Lund's district heating and cooling network to 90 000 residents in the new Science Village (Science Village, 2019), which will be built as a sustainable district from 2019, and is also studying ways to recover surplus energy and deploy solar or wind power. To find smart solutions to global challenges, and to overcome public opposition to new sustainability developments, the Bridge Summit gathers public and private stakeholders through southern Sweden's Skåne region and eastern Denmark. Recently, Lund Municipality, Kraftringen, Lunds Kommuns Fastighets and IVL Swedish Environmental Research Institute launched the CiTYFiED project to discuss sustainable district renewal (European Commission, 2017). Lund University (2018) 'found indications that a fair, transparent and open process might be more important than the distribution of project benefits' where there is public opposition. 'Perceived fairness by those affected by the change is pivotal to increasing the perceived legitimacy of transition outcomes. If this is done correctly, even contested projects can be realised. This includes that stakeholders find the space to discuss and disagree'.

4.3.5 Växjö: 'Europe's greenest city'

In 1996, Växjö became the first city in the world that aimed to become fossil fuel free by 2030. Now called 'Europe's greenest city', it is a major destination for eco-tourists and interested officials from government, universities, business and non-profit organisations that visit to study its district cooling and heating systems, energy-efficient wooden buildings, public transit run on biogas and renewables, and cycling and urban gardening in Östrabobacken's downtown. About 90% of the Växjö's energy for city heating and half of its electricity comes from trees in the form of waste branches, bark and sawdust. In 2014, Växjö emitted 2.4 million tonnes of carbon dioxide emissions per capita, compared with the EU average of 7.3 tonnes (Swedish Institute, 2016).

4.3.6 Helsingborg: body-powered heating innovations

In southern Sweden, Helsingborg has pioneered the use of 'passive houses' that generate energy from people's body heat, electrical appliances and solar panels (Swedish Institute, 2016). The Väla Gård district features solar-powered office buildings and conference rooms with air-purifying plant walls, producing more energy than it consumes and receiving the highest platinum level LEED certification for green buildings (LEED, 2019).

A similar technology is used in Stockholm, where the city's Central Station has a geothermal system capturing body heat from the quarter of a million daily commuters. The heat is transferred to water via a heat regulator, and the heated water is used to provide heating for Kungsbrohuset – a large 'eco-smart' office building. Cooling is provided by water from Lake Klara, a canal that runs through the city (Swedish Institute, 2016).

4.3.7 Linköping: carbon neutral by 2025

The home of ABB and other major aerospace and electronics companies, Linköping has a history of pro-active climate policies. In 1993, a multi-group established the Agenda 21 plan, and in 1998 the Plan of Action for a Sustainable Linköping in the Long Term. In

1997, the city built the largest biogas plant for vehicle fuel, which is produced from livestock manure, industrial food waste and slaughterhouse waste, and fuels garbage trucks, 70 public buses and 1000 other vehicles. Biogas fuels 6% of the city's vehicles. The city is a major biogas centre, with a biogas research centre at Linköping University.

Key energy supplies are provided by two CHP plants producing electricity and district heating in a highly efficient cogeneration process. About 95% of homes are connected to the district-heating network. The energy in household and commercial waste is recycled in the CHP plants and generates the majority of the heating and power produced there. In 2012, primary energy use in the district-heating system was 373 GWh compared with a figure of 1287 GWh for final energy consumption and, according to the Urban Sustainability Exchange (USE, 2019), the city has achieved a 25% reduction in carbon dioxide emissions since 1990.

In 2007, Linköping established the Environmental Technology Centre to transform environmental challenges into business opportunities. In 2011, the city adopted the goal of becoming carbon neutral by 2025. Its municipal council adopted a range of programmes for environmental development

- a long-term strategic plan for sustainable Linköping
- a general plan for physical planning, including traffic strategy
- a climate strategy, to reduce greenhouse gas emissions
- a nature conservation programme, to preserve and promote biodiversity
- an action programme for improved air quality, with a special focus on particulates
- a waste management plan based on the EU's Waste Hierarchy (USE, 2019).

Linköping continuously monitors environmental, economic and social sustainability targets and conducts a detailed follow-up of 136 indicators every 4 years. The long-term climate and environmental initiatives have had many positive effects

- the annual per capita waste has fallen from 305 to 208 kg since 1992
- waste fractions sorted at source and collected from households as paper, plastic, metal, batteries, light bulbs and electronic equipment have increased
- nitrogen and phosphorus emissions from treated wastewater have decreased by 60% and 35%, respectively, since 2001
- nitrogen dioxide concentrations in urban environments have fallen from 29 to 8 $\mu g/m^3$ of air since the late 1980s
- the number of kilometres driven per vehicle/year has decreased from 13 000 to 12 000 km over the past 10 years
- there has been a 50% increase in new 'green car' registrations
- about 56% of all the electricity used is produced locally from renewable sources and in a climate-efficient way
- there has been a 25% reduction in carbon dioxide emissions between 1990 and 2010 (USE, 2019).

The Linköping experience suggests that sustainability successes come from a combination of innovation, collaboration, political consensus and a long-term approach (Lund University, 2018).

4.3.8 Umeå: energy-efficient buildings

In northern Sweden, Umeå won the 2018 Sustainable Energy Europe Award in the Living category for building more energy-efficient buildings after a major fire destroyed the Ålidhem district in 2008. Some 400 rebuilt apartments from the 1960s and 1970s and 137 new apartments reduced the city's energy use by half using Echologs terminals for displaying and monitoring electricity, heat and water consumption (Swedish Institute, 2016).

4.4. Healthy cities for healthy people

There is, of course, a close connection between healthcare, wellbeing and improved environment. Sweden has adopted numerous SD policies to enhance its quality of life, health and wellbeing, which have guided local policy-making. In 2013, its life satisfaction was among the highest in the OECD. Between 2005 and 2011, exposure to $PM_{2.5}$ air pollution decreased sharply, particularly after 2012. Life expectancy has increased 1.5 years since 2005, reaching 82.3 years in 2015, about 2 years higher than the OECD average (OECD, 2017). In the U.S. News and World Report (2019) lists of best countries, Sweden ranked first for Green Living, for Women and Raising Kids, second for Quality of Life, and fourth for Business and Citizenship, and holds sixth place in the overall ranking.

Contrary to popular understanding, Swedish healthcare, which represented 11% of GDP in 2014 and is 83% publicly financed, is primarily paid for by local governments; 20 county councils provided 57% of the financing, 290 municipalities 25%, and the state 2%. According to the Commonwealth Fund, in 2015, healthcare accounted for 89% of county council spending (Glenngård, 2019). Healthcare is not totally free, but provided on a cost-sharing basis. Under the 1982 Health and Medical Service Act, county councils are required to provide good-quality health, medical care and dental care to all legal residents. Dental care is covered by county councils until 23 years of age, beyond which it is subsidised by the Swedish state. Municipalities are responsible for elderly care and for people with physical disabilities and psychological disorders. The National Patient Survey measures healthcare quality annually, and is coordinated by the Swedish Association of Local Authorities and Regions. In 2003, the nation implemented a nationwide strategy to improve wellbeing among the poor and unemployed. As a result, municipalities are motivated to plan and develop healthy communities and cities to reduce and minimise healthcare costs.

4.5. The future: resilient cities

Sweden has achieved its many of its early SD goals, but its policies are becoming inadequate in an era of accelerating climate change. During the summer of 2018, forest fires on a large scale struck for the first time throughout Sweden due to abnormally warm weather. Sea rise and flooding require resilience planning among its coastal towns and cities. Today, the UN warns that the world must develop smart sustainable communities (eco-towns and villages) and cities (eco-cities) to address climate goals at different levels of urban scale.

As introduced in Chapter 1, in the early years after Earth Day 1970, Richard Register and other climate activists coined the term 'eco-cities', which is now rather outdated because its idealistic concepts were top-down initiatives lacking citizen involvement, could not be integrated into existing cities because of over-simplification of urban problems, and often became isolated enclaves. Due to pressures to meet UN climate change goals, eco-city concepts are being updated (Downton, 2017) to include a broader range of pressing needs and concerns, such as

- UN Sustainable Development Goals (SDGs) to reduce climate change
- a shift from fossil fuels to renewables and electrification
- a cyclical economy and quadruple bottom-line methodologies
- disaster preparedness, resilience planning and recovery
- the increasing number of climate refugees and migrants
- broader grassroots citizens' involvement in the 'social economy'
- integration of eco-city districts with the rest of the city to avoid 'Frankenstein urbanism'
- smart cities data integration into sustainable cities
- exponential technologies (augmented/virtual reality, internet of things, artificial intelligence, blockchain, cloud computing, etc.)
- the growing public demand for active citizen engagement
- community-based healthcare, training and public services.

A major challenge to new eco-cities is to avoid creating homogeneous, isolated 'ecological enclaves' of the wealthy, totally unrelated or irrelevant to the problems of surrounding cities and towns. In the past, Sweden was relatively isolated from global urban trends, but growing immigration and climate change are strongly impacting its local economy and politics. For example, Sweden's decision to accept over 160 000 Syrian refugees in 2016 led to severe housing shortages, overloading of public services and strong populist anti-immigrant opposition.

How is Sweden dealing with these emerging challenges? How are Swedish towns and cities preparing themselves to implement new SD policies at the local level? Some of the recent initiatives being proposed by various organisations and municipalities are given below.

- **International collaboration.** The public-facing Swedish government website (www.government.se) provides information on myriad government policies and initiatives supporting smart approaches to SD. Sweden is signing numerous international agreements to co-develop clean energy technologies and tap global markets with California, France, China, India and other nations. In 2015, Sweden signed to implement the UN SDGs. The Clean Energy Ministerial and the Mission Innovation are two key initiatives to support the implementation of the Paris Agreement on climate change. In 2018, Sweden, the European Commission and the other Nordic countries hosted the 2018 Clean Energy Ministerial, to seek joint efforts to support the development and deployment of clean energy technologies. For developing regions, Sweden has outlined a strategy for global development cooperation in the areas of environmental sustainability, sustainable climate and oceans, and sustainable use of natural resources during 2018–2022, to promote the 2030 Agenda, the Addis Ababa Action Agenda and the Paris Agreement.

- **Climate policies.** In June 2017, the Swedish Parliament adopted a climate policy framework containing new climate goals, a climate act and plans for a climate policy council. By 2045, Sweden aims to have net zero emissions of greenhouse gases into the atmosphere, then negative emissions. Emissions covered by EU regulations should, by 2030, be 63% lower than emissions in 1990, and 75% lower by 2040. Emissions from domestic transport, excluding domestic aviation, will be reduced by 70% by 2030 compared with 2010. As part of its Equal by 30 initiative, Sweden has committed to globally promote equal pay, equal leadership and equal opportunities for women by 2030.
- **Resilience planning.** To prepare for climate change, Swedish organisations have promoted resilience policies and programmes for many years. The Stockholm Resilience Centre (SRC), founded in 2007, is a non-profit independent research institute specialising in SD and environmental issues that advises policy-makers on ecosystem management and long-term SD in Europe and elsewhere (SRC, 2019). The SRC works on climate change, landscapes, water resources, land use, food security, marine systems and urban systems with the aim to generate insights to shift governance and management towards sustainability. An example is the BALTICLEAD programme (formerly known as LEAP), a joint initiative between the SRC and the Swedish Institute, which is a leadership programme designed for system entrepreneurship and transformations to sustainability in the Baltic Sea region and Eastern Europe.
- **Public housing.** Sweden plans to build 700 000 public housing units and update its 1960s ‘million homes’ programme over the next decade, to address a serious housing shortage and to make houses energy efficient. The new and rebuilt housing units will enable Sweden to provide the latest smart building technologies. The goal is also greater integration with the wider society, to avoid the social isolation of current immigrants in 1960s ‘million homes’ suburbs.
- **Social enterprises.** This sector is still small and not well defined, but there is growing interest in SD and a new strategy to strengthen social enterprises and social entrepreneurship. In particular, integrating immigrants into the workforce is a major challenge, which is being addressed by social enterprises (Farmbrough, 2018). In July 2018 at Almedalen Week – the annual Swedish gathering to discuss politics, business and civil society – over 4000 seminars and panel discussions reflected growing interest in how business can improve society (Kenestad, 2018).

4.6. Lessons from Swedish SD

Sweden is unique since it has a century-long history of actively promoting SD policies at national and local levels: energy taxes since the 1920s, district heating and biofuels in the

1970s, broad energy and transportation policies in the 1990s, and formal SD government agencies in 2003, and goals of becoming carbon neutral by 2045 and completely switch to fossil-fuel-free mobility by 2050. Unlike other Nordic nations with oil fields, Sweden is vulnerable to high oil/gas prices, especially with the maturation of North Sea oil, so it must rely on local energy sources, such as timber-based biofuels, industrial and urban waste, solar power, wind, hydropower, nuclear power and heat pumps to reduce its vulnerability to politically instability among OPEC nations. Its current policies are to become self-sufficient in renewable energies in order to support its industries and cities, create low-carbon jobs, and achieve its sustainability goals under the Paris Agreement and EU policies. Sweden has an advantage, since it is wealthy, has citizen support, and is experienced in building eco-towns and eco-cities, increasingly using smart technologies.

What lessons can nations and cities learn from Sweden?

- **Long-term vision.** Developing national SD policies for all sectors of the economy is a long-term, difficult process of developing and testing complex policies and programmes. They are crucial for guiding eco-city development.
- **Tax policies.** Gas and carbon taxes can be implemented to promote energy efficiency and a shift to renewable energies, but – as France's 'yellow vests' protest against fuel tax increases has shown – they must be implemented with full citizen involvement, especially by under-represented rural areas and poor communities (Donadio and Meyer, 2018).
- **Local resources.** Renewable energies relying upon local resources, such as Sweden's vast forests, timber waste, urban waste and even commuter body heat, are the most accessible and efficient. Other nations need to tailor their energy policies accordingly. A common denominator for all nations and cities is recycled urban and industrial wastes, which can produce significant power for district heating – as Malmö and Gothenburg have done.
- **Economic revitalisation.** District heating and cooling can become the foundation for economic revitalisation since it requires smart grids and other new clean technologies. Biofuels and other renewable energies can trigger business and urban innovations with the right combination of tax policies and publicly financed transition programmes.
- **Citizen participation.** Although government plays a major role in SD policy-making, implementation is done at the local level, with increasing attention paid to active citizen engagement due to the rapidly changing nature of Swedish society, which is facing the challenges of social integration of immigrants, climate change and the decline of heavy industries.

However, the Swedish sustainability model has several limits that prevent direct applicability to other nations, as listed below.

- **Small urban population.** Sweden has a small population of 10 million people, most of whom are concentrated in cities and towns, so its policies can be focused on urban areas.

- **Rich resource base.** The nation has plentiful undeveloped land, water and energy resources (timber, hydropower, wind) so it is not environmentally stressed like many nations.
- **Homogeneous.** In the past, Sweden was monocultural, so agreement over strong sustainability policies was easier to achieve. Today, however, Sweden must deal with a more complex, multicultural society with 183 nationalities accounting for about 18% of its population. Many of the more recent incomers live in segregated 1960s towns, so social integration has become a growing issue in urban policies and planning. More specifically, recent and anticipated levels of immigration have led to strain on social resources; more sensitive public policies and new technologies are seen as a means to address aspects of this strain.
- **Governance.** Sweden has a long history of strong governance, but at the cost of high taxes like other Nordic nations, topping 61% for personal income in 2018, which is up substantially from 57% in 2016 due to high immigration and other social costs. Such high rates of tax may not be acceptable to more market-driven nations. Moreover, like other Nordic nations, Sweden ranks at the top in 'lack of corruption', so its citizens trust officials to spend their money wisely on sustainable infrastructure.
- **Political neutrality.** Sweden's industrial base was not destroyed during World War II by the Nazis due to its neutrality, which enabled it – like the USA – to recover more quickly than other European nations. During the post-war period, Sweden's neutrality and small army enabled it to divert money from military spending to education, healthcare and the environment. Nevertheless, Sweden was the 20th highest arms exporter in 2017 (Knoema, 2019) and the fifth per capita in 2014 (Jackson, 2014), so it has enjoyed peace while exporting arms. However, with growing Russian presence in the Baltic, Sweden is building closer ties with NATO, reinstated military service in 2018 and is boosting military spending (Duxbury, 2018).

The combination of large, undeveloped areas of land, rich natural resources, a small population, political neutrality, recent cultural diversity, strong governance, high taxes, a long history of sustainability policies and programmes, citizen environmental support and advanced technologies enabled Sweden to test new ideas and invest in new technologies, which is rather unique in the world. Sweden demonstrates that with long-term SD goals, policies and programmes, it is possible for a small nation to become environmentally sustainable while developing a high quality of life and strong economic growth.

Sweden is doing well today, but its future will be much more complex due to a variety of major trends: the decline of older industries, which account for over 50% of the GDP, an ageing population with higher welfare costs, immigration pressures, climate change, increasing military spending in the Baltics, and the impacts of exponential technologies, such as artificial intelligence, virtual reality, the internet of things and blockchains being adopted by Swedish companies and cities. Sweden is pursuing a proactive role by partnering with leading sustainable regions such as California and France, as well as export markets such as China and India, to prepare for a more substantial shift to renewable energies and carbon neutrality. As a small export nation without an oil/gas industry,

Swedish leaders know the nation's future depends upon energy efficiency, a shift to renewable energies, and a close collaboration with SD partners and global markets. It may be a big fish in a small Nordic pond, but Sweden will always remain a small fish in a massive global ocean.

REFERENCES

Ahlberg M (2009) Sustainable development in Sweden – a success story: discourse analysis. *L'Europe en Formation* **2(352)**: 157–179. https://www.cairn.info/revue-l-europe-en-formation-2009-2-page-157.htm (accessed 20/06/2019).

Alopaeus T (2013) *Policy instruments in the Swedish climate policy.* Department for Policy Analyses on Climate and Air Pollution, Swedish Environmental Protection Agency, Stockholm, Sweden. http://www.lvif.gov.lv/uploaded_files/sadarbiba/seapplus/T_Alopalus_Policy_instruments_Sweedish_climate_policy.pdf (accessed 20/06/2019).

Atkisson A (2006) Letter from Sweden: fossil-fuel free by 2020, maybe. *WorldChanging.* https://web.archive.org/web/20061115133219/http://www.worldchanging.com/archives/004832.html (accessed 20/06/2019).

Brittlebank W (2015) Sweden to become the world's first fossil fuel-free nation. *Climate Action.* http://www.climateaction.org/news/sweden_to_become_worlds_first_fossil_fuel_free_nation (accessed 20/06/2019).

Confederation of Swedish Enterprise View of Sustainable Development (2002) *The Confederation of Swedish Enterprise View of Sustainable Development.* Confederation of Swedish Enterprise View of Sustainable Development, Stockholm, Sweden.

Donadio R and Meyer R (2018) France's 'yellow vest' protesters aren't against climate action. *The Atlantic.* https://www.theatlantic.com/international/archive/2018/12/france-yellow-vest-climate-action/577642 (accessed 20/06/2019).

Downton P (2017) Seven things you need to know about ecocities. *The Nature of Cities.* https://www.thenatureofcities.com/2017/03/05/seven-things-need-know-ecocities (accessed 20/06/2019).

Duxbury C (2018) Under threat, Sweden rediscovers its Viking spirit. *Politico.* https://www.politico.eu/article/under-threat-sweden-rediscovers-its-viking-spirit-nato-russia (accessed 20/06/2019).

ElectriCity (2019) https://www.electricitygoteborg.se/en (accessed 20/06/2019).

Elkington J (2017) Saving the planet from ecological disaster is a $12 trillion opportunity. *Harvard Business Review.* https://hbr.org/2017/05/saving-the-planet-from-ecological-disaster-is-a-12-trillion-opportunity (accessed 20/06/2019).

European Commission (2017) European 'energy days' kick-started in Lund. https://cordis.europa.eu/news/rcn/141379/en (accessed 20/06/2019).

Farmbrough H (2018) How a Swedish social enterprise is transforming lives. *Forbes.* https://www.forbes.com/sites/heatherfarmbrough/2018/01/31/how-a-swedish-social-enterprise-is-transforming-lives/#174c9dac6aae (accessed 20/06/2019).

Fitzgerald J (2016) US city planners can learn lessons from the successes of Malmö's eco-districts. LSE US Centre. https://blogs.lse.ac.uk/usappblog/2016/03/31/us-city-planners-can-learn-lessons-from-the-successes-of-malmos-eco-districts (accessed 20/06/2019).

Glenngård A (2019) The Swedish health care system. The Commonwealth Fund, International Health Care System Profiles. https://international.commonwealthfund.org/countries/sweden (accessed 20/06/2019).

Global CCS Institute (2013) *Low Carbon Energy 2050: Local and Regional Impact and Perspectives*. Global CCS Institute, Melbourne, Australia. https://hub.globalccsinstitute.com/publications/low-carbon-energy-2050-local-and-regional-impact-and-perspectives/33-se-city-stockholms-climate-initiatives (accessed 20/06/2019).

Holmåker C (2017) Gothenburg ranks number one among sustainable and innovative cities. go:teborg&co. http://goteborgco.se/en/2017/11/gothenburg-ranks-number-one-among-sustainable-and-innovative-cities (accessed 20/06/2019).

Ignatieva ME and Berg P (2014) Hammarby Sjöstad – a new generation of sustainable urban eco-districts. *The Nature of Cities*, 12 February. https://www.thenatureofcities.com/2014/02/12/hammarby-sjostad-a-new-generation-of-sustainable-urban-eco-districts (accessed 20/06/2019).

Jackson A (2014) Peace-loving Sweden and Switzerland are among top arms exporter per capita. *PRI*. https://www.pri.org/stories/2014-05-23/peace-loving-sweden-and-switzerland-are-among-top-arms-exporters-capita-world (accessed 17/04/2019).

Jonsson NO and Berglöf C (2016) Nuclear power plants in Sweden during the last 40 years. *European Nuclear Society* **50**. https://www.euronuclear.org/e-news/e-news-50/sweden-print.htm (accessed 20/06/2019).

Kenestad T (2018) Sweden and social enterprise: making the new national strategy count. *Pioneers Post*, 12 December. https://www.pioneerspost.com/news-views/20181212/sweden-and-social-enterprise-making-the-new-national-strategy-count (accessed 20/06/2019).

Knoema (2019) Arms exports in constant prices of 1990. https://knoema.com/atlas/ranks/Arms-exports (accessed 20/06/2019).

LEED (2019) http://leed.usgbc.org/leed.html (accessed 20/06/2019).

Lund University (2018) How local communities can transition to sustainable energy systems. *Phys.org*. https://phys.org/news/2018–05-local-transition-sustainable-energy.html (accessed 20/06/2019).

McCarthy N (2019) The US cities with the worst traffic problems. *Statista*. https://www.statista.com/chart/12855/americas-most-congested-cities (accessed 20/06/2019).

Ministry of the Environment (2006) *Strategic Challenges – A Further Elaboration of the Swedish Strategy for Sustainable Development*. Ministry of the Environment, Stockholm, Sweden. Comm. 2005/06:126, p. 9. https://www.government.se/legal-documents/2006/03/comm.-200506126 (accessed 20/06/2019).

OECD (Organisation for Economic Co-operation and Development) (2014) OECD Environmental Performance Reviews: Sweden 2014. http://www.oecd.org/env/country-reviews/oecd-environmental-performance-reviews-sweden-2014–9789264213715-en.htm (accessed 20/06/2019).

OECD (2017) *How's Life in Sweden? OECD Better Life Initiative Country Note*, November. http://www.oecd.org/statistics/Better-Life-Initiative-country-note-Sweden.pdf (accessed 20/06/2019).

Reinfeldt F (2008) The New Swedish model: a reform agenda for growth and the environment. Speech given at the London School of Economics and Political Science, 26 February. https://web.archive.org/web/20080927103017/http://www.regeringen.se/sb/d/10294/a/99239 (accessed 20/06/2019).

Reportlinker (2018) Global construction outlook to 2022, Q3 2018 update. *Cision PR Newswire*, 8 October. https://www.prnewswire.com/news-releases/global-construction-outlook-to-2022-q3–2018-update-300727015.html (accessed 20/06/2019).

Sahlin M (2015) Bränslen. *Energimydigheten.* https://www.energimyndigheten.se/statistik/branslen (accessed 20/06/2019).

Savage M (2017) All eyes on Malmö, but not because of Trump. BBC. http://www.bbc.com/capital/story/20170306-all-eyes-on-Malmö-but-not-because-of-trump (accessed 20/06/2019).

Science Village (2019) https://sciencevillage.com/en (accessed 20/06/2019).

Stockholm Convention (2001) Overview. Stockholm Convention on Persistent Organic Pollutants. http://chm.pops.int/TheConvention/Overview/tabid/3351/Default.aspx (accessed 20/06/2019)

SRC (Stockholm Resilience Centre) (2019) https://stockholmresilience.org (accessed 20/06/2019).

Svebio (2017) Biofuels for transport. https://www.svebio.se/en/about-bioenergy/biodrivmedel/ (accessed 28/08/2019).

Swedish Institute (2016) 7 examples of sustainability in Sweden. https://sweden.se/nature/7-examples-of-sustainability-in-sweden (accessed 20/06/2019).

Swedish Institute (2019) Sweden tackles climate change. https://sweden.se/nature/sweden-tackles-climate-change/ (accessed 20/06/2019).

Torres N (2017) How data and cities can shape the future of mobility. *Data-Smart City Solutions.* https://datasmart.ash.harvard.edu/news/article/how-data-and-cities-can-shape-the-future-of-mobility-1182 (accessed 20/06/2019).

U.S. News and World Report (2019) Best countries for green living. U.S. News and World Report. https://www.usnews.com/news/best-countries/best-green-living (accessed 20/06/2019).

USE (Urban Sustainability Exchange) (2019) Linköping carbon neutral 2025. https://use.metropolis.org/case-studies/linkoping-carbon-neutral-2025 (accessed 20/06/2019).

Wilde B (2016) How Sweden became the world's most sustainable country: top 5 reasons. ADEC Innovations Blog. https://info.esg.adec-innovations.com/blog/how-sweden-became-the-worlds-most-sustainable-country-top-5-reasons (accessed 20/06/2019).

Part 2. Mobility

ICE Publishing

Nuttall, William J, Gibson, David V, Trzmielak, Dariusz and Ibarra-Yunez, Alejandro
ISBN 978-0-7277-6425-6
https://doi.org/10.1680/emsc.64256.083

Chapter 5
Autonomous vehicles and the urban mobility ecosystem

Stephen Potter, Alan-Miguel Valdez and Matthew Cook
The Open University, Milton Keynes, UK

5.1. Introduction

The rapid development of autonomous vehicle technologies by major information technology (IT) companies as well as established car manufacturers is set to be a major disruptive innovation in a sector that, for the past 80 years, has been characterised by incremental shifts in technologies and behaviours. Indeed, developments in autonomous vehicle technologies are now moving so fast that any detailed snapshot of trials, proposals and projects would soon be out of date. But what is more, as noted by Lyons (2016), the development of autonomous vehicle technology is just one of several socio-technical factors that together are changing the urban mobility regime. Increasingly, there is a sense that our mobility system is entering a time of transition, the outcome of which is very uncertain. As such, this chapter seeks to provide a strategic view of the key issues, opportunities and challenges presented by the emergence of autonomous vehicles and how these interact with other major trends and changes in the urban mobility regime (Figure 5.1). Using a case study of public transport systems, this chapter explores how autonomous vehicle technologies could become part of the solution to the problems of our urban mobility ecosystem, rather than a missed opportunity or something that worsens the situation.

5.2. The slow-burning urban transport crisis

The main issues around the slow-burning, intractable crisis facing our existing urban transport ecosystems are well known. Transport is a vital part of all economies and societies, yet complex issues with multiple interactive causes have combined to produce a range of serious problems. These include congestion of infrastructure, dangerous levels of local pollutants (with key emission of carbon monoxide (CO), nitrogen oxides (NO_x), particulates and sulphur dioxide (SO_2) coming from transport sources), accident casualties and a variety of negative social impacts (such as unhealthy sentient lifestyles) as well as the major issue of carbon dioxide (CO_2) and other greenhouse gas emissions contributing to global warming (Banister, 2005).

The issue of global warming provides a good illustration of how intractable the situation is for the transport sector. As part of its international obligation to cut emissions

Figure 5.1 A trial autonomous 'pod' outside Milton Keynes station. (Source: Transport Systems Catapult)

contributing to climate change, the UK's 2008 Climate Change Act seeks an 80% cut in CO_2 emission by 2050 relative to the base year of 1990. This requires a 3% CO_2 reduction per annum and a major reduction in CO_2 from the transport sector by 2020. Measures to improve car fuel economy appear to have made the sort of progress needed to address the climate change targets. In the UK, between 2000 and 2015 the average fuel consumption of petrol cars improved by 32% and that of diesel cars by 28%. More recently, there have been substantial initiatives to support the introduction of low-carbon hybrid and electric vehicles (Rietmann and Lieven, 2019).

Yet, although from 1990 to 2012 the UK's end-user greenhouse gas emissions dropped by 26%, those from transport were reduced by only 4% (Department of Energy and Climate Change, 2014). As a result, transport's share of emissions has grown from 18% of the total to over 23%. Rather than the required 3% CO_2 reduction per annum, the transport sector has not even managed a 3% reduction *per decade*. The inadequate reduction in Britain's CO_2 emissions is actually better than in most comparable developed economies. Across Europe as a whole, in 2015 transport emissions were 26% *above* 1990 levels (European Environment Agency, 2018) and globally the growth in transport's CO_2 emissions is even higher. The International Energy Agency (2019) reported that, in 2016, transport's global CO_2 emissions were 71% higher than in 1990 and accounted for one-quarter of total CO_2 emissions. As noted by Frey (2018) among others, fuel economy improvements have been swallowed up by wider socio-economic factors increasing transport dependency.

So, when considering the potential impact of autonomous vehicle technologies on urban mobility, it is important to examine the overall travel-generating ecosystem. Thus, the key question is how might autonomous vehicles affect the urban mobility ecosystem as a whole. Will the development of autonomous vehicles merely be another factor worsening transport's social and environmental impacts, or might they (if applied appropriately) be part of a shift towards improving the situation?

5.2.1 UK transport policies and the urban mobility ecosystem

One way to start exploring what could be the impact of autonomous vehicles is to first explore recent urban mobility trends and policy responses. In the UK, as well as in a number of other developed economies, public policy has sought to shift trips from the car to public transport systems in order to address urban congestion, social and environmental issues. This has been done largely through regulation, the use of public funds being devoted to infrastructure investments and ongoing public transport fare subsidies. For example, the 25 km Cambridge Guided Busway cost £180 million to build, and London's Crossrail megaproject has cost some £17 billion. In London alone, subsidises to public transport fares are £3 billion per annum.

Such policy responses frequently focus upon commuting to and from workplaces, but although peak hour traffic is noticeable for the congestion and frustration it causes, other travel purposes are actually far more substantial. Indeed, it is changes in overall travel behaviour that helps to explain why the transport sector has failed to cut its greenhouse gas emissions, why congestion has worsened and air quality issues persist.

Travel behaviour is driven by deep-rooted economic and social practices that have led to travel becoming increasingly dispersed in time, space, across functions and across demographics. Transport planning focuses upon workplace journeys, but these now constitute only 15% of all trips in the UK and only 3% of business trips (Department for Transport, 2016). As noted in the report *All Change?* (Marsden *et al.*, 2018), the major areas of travel growth are now in leisure purposes (which have grown to 31% of all trips), shopping (19%, although internet shopping has led to some decline of late) and highly dispersed personal business trips (19%). In space, the strongest growth has not been along major corridors to city centres but in suburban, urban fringe ('peri-urban') and rural areas.

The rise in car ownership and use has much to do with this dispersal of travel demand, but it is also a result of fundamental shifts in our economy and society. For example, employment is less stable than in the past, which, together with increasing housing costs, leads to longer commuting distances, home working and weekly commuting. The peak time for motorway traffic around London is now Friday afternoons and Sunday evenings. Social networks are also increasingly fluid – they have changed as our 24/7 society has developed and, despite the growth in social media (or possibly because of it), people travel more and further for social purposes.

The change in travel by demography was noted in the 2016 report *On the Move* (Headicar and Stokes, 2016) and the report *All Change?* (Marsden *et al.*, 2018). For the elderly, driving licence ownership has markedly increased, but among 'millennials' (up to 34 years in age) it dropped from 80% in 1995 to 60% in 2016. It appears that the young, IT-savvy millennials are buying into new behaviours, with car ownership being less of a priority compared with other lifestyle aspects. They use urban public transport and emerging IT-based cab services such as Uber.

Smart car and smart city technologies will accelerate and redirect changes in the urban mobility ecosystem, but in themselves will not be the sole determinate. A range of emergent digital technologies, often referred to as the 'internet of everything', are interacting to provide 'the most significant technological trend faced globally' (Leviäkangas, 2016). These include instrumented sensor networks, mobile devices and consumer electronics, especially combined with internationally integrated telecommunication infrastructure and advanced programming algorithm methodologies. These not only affect transport technologies and methods but are instigating more fundamental structural changes in the nature of employment, leisure, social activities, shopping and so on. We are just seeing the beginnings of how these interactions are taking place, and it is very difficult to envisage even the broad direction this will eventually take, including a baseline question of whether these developments will assist replacements for or be complements to travel.

One danger is of autonomous vehicle developments being focused upon past travel patterns and behaviours, and there are indications of this in the way they are presented as a 'solution' to commuting or to improve already outdated business models. Generally speaking, new mobilities technologies and services need to find viable roles within existing mobility patterns and practices. From such niches within the existing transport regime they can develop and move to the mainstream. But the new technologies and services will also need to adapt to rapidly changing digital-driven trends. Smart cities will lead to a whole range of new behavioural patterns, and urban space itself will change. Some commentators envisage radical changes in travel behaviour, even speculating the end of car ownership, as autonomous vehicle services could provide a better level of service at a lower cost.

Research by Davidson and Spinoulas (2015) indicated that the widespread use of autonomous vehicles would reduce parking requirements, but that they would increase trip making and trip length as car use becomes easier, cheaper and less stressful. That would have structural effects on city form and activity locations. They also consider that the relative attractiveness of public transport would be altered. Easier car use would make public transport less attractive, even were the new technologies applied to improve the performance of metro, bus and rail services.

But Davidson and Spinoulas (2015) also suggested that there is the potential for autonomy to make public transport more attractive. New models of public transport could emerge, providing a more responsive and affordable service (as we discuss further on). That would require the redesign of the public transport service as a whole, not just using new technologies to enhance the existing way public transport is provided. The extent to which such innovations require changes to urban infrastructures will be considered in Chapter 6.

We are in a situation of uncertainty, where interactions between smart transport configurations and a wide range of emergent smart city regimes could produce a wide mix of potential outcomes. It is possible that resultant urban mobility behaviours could move towards a reduction in the environmental and social impacts of transport, while other outcomes would lead towards worsening congestion and environmental impacts. As noted by Lyons (2016), the level of uncertainty involved is substantial, with there being not just gaps in knowledge but a divergence of views on both outcomes and policy responses all playing out amidst an extremely complex web of socio-technical relationships.

5.2.2 The socio-technical configuration of public transport

The need for a regime-level understanding can be explored by focusing upon the example of urban public transport and the way in which it is viewed by the transport planning system. A concentration on commuting trips along busy corridors from suburbs into city centres leads to actions that address this situation well, such as building metro or tram lines and improving bus services. Urban planning policies have followed this corridor system design, with settlements and facilities clustered along high-density corridors radiating from city centres. Smart city systems and technologies are increasingly applied to improve such traditional corridor services, for example the development of e-ticketing and informational apps (Figure 5.2). But with the vast bulk (around 80%) of personal travel now not following a concentrated transport corridor pattern, this focus on late 20th century thinking is increasingly out of step with 21st century realities.

Figure 5.2 A travel information app. (Source: MK:Smart)

Consider the picture of a bus shown in Figure 5.3. It is quite easy to identify the technological changes in bus design since the first buses came into use some 200 years ago. A user from the 19th century would hardly recognise a modern bus. But in terms of a bus service system, that 19th-century traveller would be entirely at home because the system design is essentially the same now as it was then. Buses operate to a timetable on fixed

Figure 5.3 A 19th-century horse bus. The technology used has vastly changed, but not the system design

routes. Passengers walk to bus stops, signal a bus to stop and pay a fare. Then they signal the bus to halt at a bus stop, and walk to their final destination or change to another route if that is needed. Details of how you obtain information on bus services and payment may now be electronic, but the basics of the service design are essentially the same.

It is perhaps remarkable that this system configuration has remained unaltered for 200 years, even though the technology of public transport vehicles, fare collection and any associated track and infrastructure have seen considerable technical developments (Daganzo, 2010).

Smart and autonomous vehicle technologies could be applied to further improve this existing model. Indeed, there are some remarkable developments emerging. Modern urban tram systems (Light Rapid Transit or LRT) have emerged as a clean and attractive public transport system for many cities, but they are expensive to build, and inflexible because their routes are fixed along their tracks. Guided buses (such as in Cambridge, Eindhoven and Adelaide) have developed, but still depend on diesel bus technology and generally do not provide a step-change quality of service.

Exploiting emerging smart technologies to the full could change this situation. A combination of inductive charging and electronic guidance means that enhanced-quality battery electric vehicles can now fulfil the function of trams without the need for costly

Figure 5.4 The Zhuzhou ART system being trialled in 2017. (Source: CRRC Zhuzhou)

and inflexible track and overhead line infrastructure. This considerably reduces the cost and time taken to introduce a high-quality public transport system. It is also a design that is amenable to autonomous operations as technology advances; it could be applied partially (e.g. first on segregated sections) and then, as autonomous driving in mixed traffic is proven, more generally.

While city authorities in Europe and their mainstream providers are showing little shift from traditional tracked and powered-tram LRT systems, the Chinese have started to work through the logical implications of new smart technologies for urban public transport designs. Implemented remarkably quickly, an Automated Rapid Transit (ART) service entered trials in the city of Zhuzhou in the autumn of 2017 (Newman, 2018) (Figure 5.4). These battery-powered electric vehicles are inductively charged every 3 km at station stops, and use optical, GPS and Lidar guidance rather than rails. At present the vehicles are manually driven, but they are designed to be upgraded to autonomous operation as the technology becomes available.

Is this the direction and vision of how smart and autonomous technologies could enhance urban public transport? But let us return to that picture of a 19th-century horse bus. ART is on the direct lineage of that 1820s system design. It is still about big vehicles operating corridor routes addressing mass commuting into and out of city centres. In no way does such an approach address the point that it is the traditional public transport system itself that is ill suited to the 21st-century landscape of travel demand.

The question that is intuitively avoided is whether much of the difficulty in promoting public transport is about the suitability of the basic service design for the transport needs of people in the 21st century. While the product-level design of the service can be improved, is the key to the problem the service system design itself? Much of the reluctance to use public transport is about its core system design being unsuited to modern travel needs. It is inadequate for 21st-century patterns of travel demand, which, as noted above, are dispersed geographically across time and across functions. So, although upgrading corridor systems using autonomous and other smart technologies has potential, this will only ever have a limited impact in addressing the challenges facing urban transport.

What public transport needs is a systems-level regime change solution.

5.3. Towards a new systems model?

Symptomatic of the inability of conventional public transport to serve dispersed city travel has been the emergence of new systems that actually do this, such as city bike and car schemes as well as the growth in the use of personal local aids such as scooters, motorised skateboards and other battery-powered low-speed vehicles.

But what about the main public transport system itself? In an article written shortly before his death, Sir Peter Hall (2013) reviewed the need for a new form of public transport to effectively serve decentralised and dispersed travel demands, seeking what he called the 'Heineken' system. This term was a reference to the old Heineken lager advertisement, with Hall seeking a public transport system that 'refreshes the parts other transport cannot reach'. But Hall could not find such a design. Despite his system-level specification, Hall restricted his consideration to public transport modes conforming to the existing big vehicle/corridor system design.

There have been, however, several examples of non-corridor systems that are variedly called 'paratransit' (a term originating in the USA) or 'personalised rapid transit' (PRT). A long-established niche market example are the USA airport shuttles that take people from an airport to destinations in a particular area, dropping them off at a number of points or (for a higher fare) at a particular address. Some European airports also licence similar shuttle systems.

This is one of a number of service models explored by Enoch *et al.* (2004), who proposed a fourfold service model typology

- **Interchange services** – feeder services into higher frequency bus and rail-based corridor services. The Chiltern taxibus in the UK and Netherlands Traintaxi are examples of this.
- **Network services** – which supplement conventional corridor services in a particular place or at certain times. An example is the Helsinki Kutsuplus, a nine-seater minibus that can be ordered to a pick-up point at a certain time. There can be other passengers on board, picked up and dropped en route. An algorithm calculates the most efficient route for drop off for everyone.

- **Destination-specific services** – which include the airport shuttles that operate to most major USA airports. Workplace vanpools are another example.
- **Substitute services** – which effectively reinvent public transport by replacing conventional public transport rather than complementing it. One of the best-known examples of this type is the Taxibus scheme in Rimouski, Quebec, which in 1993 saw the city authorities replace a stage bus network with a shared taxi operation for services to suburban areas (Trudel, 1998).

A crucial weakness has been that many of these systems have been developed within the culture, structures and governance systems of the conventional public transport regime. This has resulted in them tending to be expensive to operate and generating low revenue. They have thus struggled to establish a role within the existing institutional frameworks, although in some places with a high-subsidy culture for public transport there are examples where such systems have been sustained.

The addition of autonomous vehicle technology has led to a resurgence of interest in PRT systems. In 2011, tracked four-seater autonomous vehicles (Figure 5.5) began operating from Heathrow Airport's Terminal 5 to its main car park, replacing large buses. These operate on entirely segregated tracks and, although the design of the Heathrow PRT has been overtaken by more-advanced technologies, this does represent an initial niche implementation of a different public transport system – one that is suited to the 21st-century transport ecosystem.

The service design concept is not to run along a corridor route, like buses and trains, but to operate individual journeys across a network of segregated tracks using small vehicles

Figure 5.5 Heathrow's autonomous Ultra tracked pod. (Source: Stephen Potter)

that take people directly from their origin stop to their destination stop. The battery-electric 'pods' wait for customers at local stops, and when one pod is occupied, another automatically replaces it to await the next customer. The key characteristics of the service design are radically different to conventional public transport. You don't wait for a vehicle – it waits for you; the vehicle is managed through mobile devices – the passenger has control with a service built around them and their needs. This is an example of the way that autonomy could transform public transport to become the more responsive and affordable system sought by Davidson and Spinoulas (2015).

5.3.1 Analogue versus digital wars

At the same time as tentative developments have taken place in segregated autonomous PRT systems there has been rapid progress in another transition path towards a new urban public transport system. This has been the appearance in the taxi and private hire (minicab) market by new businesses from the digital economy world.

As travel has become increasingly dispersed, the only existing public transport mode that could fit this pattern has been the taxi. However, the taxi is a mode of transport with a system design legacy of its own, built around a 19th-century business model of street hailing, taxi ranks and high fares. But, using new IT tools, organisations such as Uber and Lyft have developed new service models that cut costs and fares (Boeckel *et al.*, 2012), thus challenging the incumbent 19th-century taxi model. Uber's lean model involves a user-friendly booking and payment app, crowdsourced owner-drivers, highly efficient scheduling and back-office software. Together, this package can outperform incumbent minicab operators on price and performance, and leaves the old taxi model standing (if not redundant).

The rapid appearance of this sector invader has invoked the politically powerful wrath of the taxi industry in cities around the world (Figure 5.6). A number of city authorities have supported the existing cab regime by constraining or banning Uber, and in at least one case have even jailed their executives (Groden, 2015). Yet, despite its defects, the Uber-style model is a configuration that matches the 21st-century mobility landscape; in an analogue versus digital war, the analogue system will eventually give way to that of the digital invader. Indeed, many conventional minicab operators in the UK (and even conventional taxi operators) are adopting app-based booking and management systems to optimise vehicle operations, cut costs and increase competitiveness.

Adding to this lean business model, operations such as ViaVan have moved towards a full PRT service. ViaVan uses shared eight-seater vehicles to provide point-to-point services within a defined service area (ViaVan Technologies, 2019). Its business model includes flexible vehicle rental for drivers, cashless payments, app booking and scheduling software to ensure a driver's income. This combination and integration of supporting technologies results in customers paying only a bus-level fare for a point-to-point journey. ViaVan fares within the Milton Keynes service area are about mid way between bus and cab fares. As in the case of ART in Zhuzhou, those in the forefront of new digital cab systems are looking to a transition to autonomous operations. If the shared ViaVan service with a driver can approach bus fare levels, an autonomous version could

Figure 5.6 Budapest taxi drivers protest against Uber, 2016. (Source: Elekes Andor, licensed under CC BY-SA-4.0)

reduce the cost of urban public transport by an order of magnitude. And that would be for a 24/7 service of a consistent quality.

Behind the digital transport service invaders is a global socio-technical mega-trend that will be hard to stop, even if it is held back a while in some places. Big-data IT systems now allow passengers and service providers to communicate directly with each other, and trip demands and available transport supply can be matched or brokered almost instantaneously. In many other sectors (music regarding downloads and tourism with customised websites being but two of many examples) there has been a readjustment between incumbent and new technology. Local transport services are only just starting to be touched by the regime-restructuring power of the digital revolution. In the long term, for some locations (particularly peri-urban and rural) and time periods (evenings and night-time) Uber- or ViaVan-style systems and their autonomous successors are set to become the public transport of the digital age.

The potentially transformative impact of 'small autonomous vehicle/small infrastructure' public transport is that, rather than people needing to adjust their behaviour to the schedules and routes of a bus or metro, they can travel directly, whenever they want, on services that can operate 24 hours a day, 7 days a week. Crucially, the level of service can be maintained at times of low demand, thus overcoming the poor quality of infrequent evening, night and Sunday services experienced today. This is a service system design that matches the socio-economic culture of the 21st-century city – not one requiring 21st-century society to conform to a 19th-century transport system architecture.

Small-vehicle, on-demand, public transport systems could cut energy use through operating vehicles only when needed at a good vehicle occupancy. This energy and emission reduction from such service optimisation impacts would be greatly enhanced should the service quality lead to a lowering of car dependence. Given the trend of millennials to defer car ownership in present cities, such a potential could be considerable.

A shift to a public transport service system of this type has major implications for transport and urban planning. Although conventional-corridor big-vehicle systems can be expected to continue to serve the market for which they are suited, a small-vehicle non-corridor system can provide a viable alternative to private car use in suburban and urban fringe situations. For public authorities, the infrastructure costs are relatively low and possibly could be subsumed within a commercially viable package. It is this system-level regime change that has the potential to deliver energy and sustainability gains together with greater social inclusion and economic benefits, such that the aims of 'smart city' development parallels 'eco-city' development.

We are possibly in a transitional phase in which several threads are combining to produce a new intermeshed system of public transport services that have the potential to deliver a more-sustainable pattern of travel behaviour. Quite how these threads will combine (and what others may come into the melee) is as yet unclear. An ideal might be seeking a vision that would include high-capacity rail-based systems along heavily trafficked corridors, coupled with other medium-density routes operated by ART-type vehicles. These would link in with PRT-type autonomous vehicle services as well as shared bike and other personal transport vehicles. Together this could produce a service system suited to 21st-century urban mobility needs.

The big question that remains is will such a positive outcome actually happen. At the moment, there seems to be a rather uncoordinated mix in which a large number of technical and trial developments are emerging from a range of actors interacting (sometimes in a friendly and sometimes a hostile manner) with established providers, legislators, labour organisations, users and so on. In the midst of all this there appears to be no real sense or vision of what is being sought from an autonomous transport future. Public policy structures have not previously had to consider such systems-level choices, with transport being a sector that has only incrementally changed over long periods of time. As such, public policy and public investment have tended to continue the traditional approach and simply sought to support short-term objectives (e.g. transfer to cleaner transport fuels) rather than identifying the strategic trajectories that offer the greatest societal and economic benefits. There is an enormous opportunity emerging that should not be missed.

5.4. Themes and issues

This chapter is not intended to reach a neat conclusion, but to use the example of public transport to identify and begin to explore the themes and issues involved as competing views and visions of an autonomous vehicle future unfold.

A key point that this public transport case study highlights is that the introduction of autonomous driving technology should not be viewed in isolation. It is not a matter of everything else being the same with autonomous driving added. Autonomous driving is one of a series of changes that are taking place – some are technological, some economic, some social and some organisational. The mobility ecosystem is a complex blend of these factors; none are fixed, and some are changing very rapidly.

In the immediate term, for public transport it is not autonomous driving that will begin a transformation. That has already begun with the innovative IT-based restructuring of the taxi model instigated by Uber and Lyft and increasingly picked up by minicab and conventional taxi operators. This sets a path towards autonomous driving as and when it becomes viable. Shared-economy developments of this model, such as ViaVan, are following a similar trajectory. For the ART-style trackless tram, it is the use of inductive opportunity charging and advanced battery technologies developed for private cars that produces the considerable benefits of lower infrastructure cost and speed of construction. In a UK context, an innovative institutional structure to manage financial risk in introducing inductive charged electric buses has also featured (Miles and Potter, 2014).

Again, several elements of the mobility ecosystem need to be combined – it is not just that one clever technical fix delivers a solution. ART and other electric vehicle systems are developments that would be further enhanced by autonomous driving (yielding an additional reduction in costs and permitting 24/7 operations), and for public transport there is the opportunity for the early introduction of autonomous driving on segregated sections.

But the public transport case also demonstrates the limits of developing within an existing system design. The cost reductions, service enhancements and network extensions achieved by a public transport service such as ART might extend corridor services from being an option from 20% of travel to 30% or maybe 40%. But it is only with a system-level change, to where a service design encompasses 80–90% of urban travel, that real benefits will accrue. And it is that final step change to a new service system design that involves autonomous driving.

The same principles apply to other parts of the urban mobility ecosystem. Autonomous driving within the owner-driver model for cars will yield limited energy efficiency improvements, which could easily be negated by autonomy generating additional travel, congestion and energy use (as noted by, among others, Davidson and Spinoulas, 2015). Such a negative outcome would be a parallel to the case of the recent improvements in individual vehicle fuel consumption being counteracted by increases in use.

So, a key theme is that, although there is a real potential for smart and autonomous vehicle technologies to deliver an attractive sustainable transport system, this potential will be unrealised if the technology is only applied within the transport system configurations that have built up around 'dumb' technologies. Applying smart technologies within the existing systems and structures will not lead to a transformation. It will lock us into unsustainability, and the other transport challenges and problems will persist.

A number of new system configurations could arise, and it seems unlikely that a solution would involve the total replacement of private car ownership and corridor public transport. A more-diverse situation might be a blend of these with new mobility services and systems that meet particular access needs, groups of people and functions.

Most approaches to autonomous vehicle systems have focused upon developing smart car and smart mobility technologies as if these in themselves represent the answer to transport's social, economic and environmental problems. How autonomy links into the changing mobility ecosystem and its institutional structures is a vital, but much neglected, area. As Kane and Whitehead (2018) point out, it is very easy to imagine emerging transport disruptions leading to 'a series of nightmare scenarios and poorer transport systems unless we have sensible and informed public policy to avoid this'. Smart cars require smart public policy. The financial, social and business model transition paths need as much (if not more) attention than technology development. There is a crucial role here for a public-authority-led process in partnership with commercial and other actors. Transport policy has developed over the past 100 years to function as a simple 'tax and spend' approach. The state uses taxation income to build transport infrastructure, complemented by other actions such as regulation measures and public transport fare subsidies. Around this approach has developed a whole state regime of organising and managing these activities. Such an approach has also been used for more recent smart transport developments, yet this system in itself is becoming less appropriate to represent public interest in our emerging digital world.

What may be needed is a less-controlling approach to governance at both national and local levels to manage the smart transport transition for the public benefit. Instead of controlling transport projects, smart transport policy is more about network building, identifying approaches that fulfil policy needs and enabling actors to deliver these. This is a very different role from the way that transport planning is presently undertaken. Public policy needs to engage and develop more of an enabling partnership approach with the new and increasingly powerful digital actors. New skills and understanding are required of the public sector. As well as smart transport needing new transport systems to realise its potential for good, it also requires a new transport policy system.

The big IT corporations understand how to blend such technical, management and network-building skills. There is a real danger that if this this key transition management role is not understood and developed by government, then, by default, the strategic direction of our transport systems will be left solely in the hands of already powerful transglobal forces.

REFERENCES

Banister D (2005) *Unsustainable Transport*. Taylor and Francis, London, UK.

Boeckel M, Sprunger B, Smith K and Work E (2012) *Uber – How a Technology Firm is Changing the Traditional Transportation Model*. Kellogg School of Management, Northwestern University, Evanston, IL, USA.

Daganzo CF (2010) *Public Transportation Systems: Basic Principles of System Design*. Institute of Transport Studies, University of California Berkeley, Berkeley, CA, USA.

Davidson P and Spinoulas A (2015) Autonomous vehicles – what could this mean for the future of transport? *Australian Institute of Traffic Planning and Management Annual Conference*, Toombul, Queensland, Australia. http://transposition.com.au/papers/AutonomousVehicles.pdf (accessed 20/06/2019).

Department for Transport (2016) *National Travel Survey 2016*. Department for Transport, London, UK.

Department of Energy and Climate Change (2014) *Statistical Release: 2013 UK Greenhouse Gas Emissions*. Department of Energy and Climate Change, London, UK.

Enoch M, Potter S, Parkhurst G and Smith M (2004) *INTERMODE: Innovations in Demand Responsive Transport*. Department for Transport and Greater Manchester Passenger Transport Executive, Manchester, UK.

European Environment Agency (2018) *Greenhouse Gas Emissions from Transport*. European Environment Agency, Copenhagen, Denmark.

Frey HC (2018) Trends in onroad transportation energy and emissions. *Journal of the Air & Waste Management Association* **68(6)**: 514–563.

Groden C (2015) Two Uber executives arrested in France. Fortune.com. http://fortune.com/2015/06/29/uber-executives-arrested (accessed 20/06/2019).

Hall P (2013) Refreshing the parts that other transport cannot reach. *Town and Country Planning*, March, 121–132.

Headicar P and Stokes G (2016) *On the Move*. Independent Transport Commission, London, UK.

International Energy Agency (2019) CO_2 emissions statistics. International Energy Agency, Paris, France. https://www.iea.org/statistics/co2emissions (accessed 20/06/2019).

Kane M and Whitehead J (2018) How to ride transport disruption – a sustainable framework for future urban mobility. *Australian Planner* **54(3)**: 177–185.

Leviäkangas P (2016) Digitalisation of Finland's transport sector. *Technology in Society* **47**: 1–15.

Lyons G (2016) Transport analysis in an uncertain world. *Transport Review* **36(5)**: 553–557.

Marsden G, Dales J, Jones P, Seagriff E and Spurling N (2018) *All Change? The Future of Travel Demand and the Implications for Policy and Planning*. Commission on Transport Demand, London, UK.

Miles J and Potter S (2014) Developing a viable electric bus service: the Milton Keynes demonstration project. *Research in Transportation Economics* **48**: 357–363.

Newman P (2018) Why trackless trams are ready to replace light rail. *EconoTimes*. https://www.econotimes.com/Why-trackless-trams-are-ready-to-replace-light-rail-1433854 (accessed 20/06/2019).

Rietmann N and Lieven T (2019) A comparison of policy measures promoting electric vehicles in 20 countries. In: *The Governance of Smart Transportation Systems* (Finger M and Audouin M (eds)). Springer, Berlin, Germany.

Trudel M (1998) The taxi as transit mode. The city of Rimouski demonstration with the taxibus. *Annual Conference of the International Association of Transportation Regulators* (IATR), Miami, FL, USA.

ViaVan Technologies (2019) What is ViaVan? https://support.viavan.com/hc/en-us/articles/360003169292-What-is-ViaVan- (accessed 20/06/2019).

ICE Publishing

Nuttall, William J, Gibson, David V, Trzmielak, Dariusz and Ibarra-Yunez, Alejandro
ISBN 978-0-7277-6425-6
https://doi.org/10.1680/emsc.64256.099

Chapter 6
Innovation in transport: intelligent transport system developments

Dariusz M Trzmielak and Remigiusz Kozłowski
University of Łódź, Poland

6.1. Introduction

The initial acceleration of a country's industries relies on being able to access appropriate transport infrastructure. In this chapter we examine the extent to which innovative cities require new urban infrastructures, and the nature of these requirements. As discussed elsewhere in this book, smart cities are fundamentally communities of people rather than a new urban machine to be optimised. While that is true, it is clear that the smart city does require new engineered infrastructures if it is to meet the challenges ahead. For example, transport expansion is seen as necessary to relieve pressure on, and to provide opportunities for, future cooperation between states, regions, cities and entrepreneurs.

Beginning with a brief overview of innovation trends in the transport sector, the chapter introduces innovation examples related to different forms of transport. The chapter then moves on to examine the practicalities and importance of an intelligent transport system (ITS) for future urban and metropolitan growth and to decrease the negative environmental impacts of the transportation sector, using the example of the implementation and analysis of the ITS deployed in the city of Łódź, Poland.

6.2. Developing innovation trends in the transport sector

In recent decades, transport has become one of the most dynamic sectors, with influence on many aspects of economic activity and on human life generally. The implementation of improved transport technologies, products and services is vital to reduce both the time and spatial disproportions between market demand and supply (Kauf, 2012). It also has a direct impact on the environment and on people's everyday lives.

Transport development has experienced rapid and innovative growth. Macro-trends include smart technologies advancing an increase in efficient urban transport and regional transport systems; the growing number of travellers (notably including ageing populations); the changing pattern in the mobility of people; continuing urban sprawl; the desire to deploy new vehicle fuels (and changes in fuel delivery technology); and new transport infrastructures. Will the system meet capacity, and, indeed, should this be the goal of system improvement? Taken together, an examination of such issues can give an insight into what will shape transport innovation in the future.

Innovative technology development (smart and green), in the form of both products and services, is a leading goal of European policy. Innovations in transport, linked to a desire to see the roll-out of green technologies (products and services), are also being promoted by urban and regional governments. This promotion can take the form of strategic and operational planning, 'green' programmes and local policies (Van Geenhuizen and Schoonman, 2010).

Typically, an efficient urban transport system is a principal policy goal for enhanced wealth creation. The public is increasingly becoming aware of energy consumption and the level of pollution produced by, for instance, vehicles using fossil fuels. Increased road infrastructure investments, often financed through higher fuel taxes, strengthen both the competitiveness and the density of public transport provision. Better transport infrastructures, such as road length and density, as well as railway length and density, have been shown to have a beneficial impact on the overall cost and performance of transport (Dobranskyte-Niskota *et al.*, 2009). New and improved vehicle fuels become more desirable as traditional, petroleum-based fuels become increasingly scarce and more costly. Furthermore, as the mobility of people and therefore the demand for transport solutions increase, existing legacy transport infrastructures create challenges for transport companies. So, one consideration is the relationship between infrastructure capacity and the growing number of travellers. This can have knock-on consequences for tourist services and for information providers and government at urban, regional and national levels (International Transport Forum, 2016). Irwin (2000) suggested that technological innovations have expanded the array of delivery mechanisms for transport. For example, for freight transport, containerisation, bulk shipping, online monitoring and online information access have cut loading times and expanded the scope of global transport and logistics.

The transport sector has such an important role in human life that it can motivate invention and generate new technology developments. Transport is not just a technology sector – it is an area of social innovation. As introduced in Chapters 2 and 5, the 'usership' of transport vehicles gains ground over ownership in meeting social needs. Social needs play a crucial role, for example, in an ageing population that faces age-related physical and psychological constraints (EMTA, 2017). Another major challenge in transport innovation is the changing pattern of urbanisation. One frequently sees city populations moving away from downtown and central urban areas into (often) newly created, lower-density suburban communities. Without sufficient public transport, populations living outside central urban areas often become car dependent. 'Urban sprawl' can lead to concentrations of private vehicles seeking to access city centres at peak times (see Chapter 5); it can also generate demands for better, publicly available, and affordable technologies and services in transport underpinned by improved transport infrastructure.

6.2.1 Innovation in transport

The various transport modes, which include road, rail, air and maritime, and their respective infrastructures, differ in terms of how innovations are developed and brought to market. The new technologies can involve decarbonising transport. New technologies and services can support or even replace transport provision. Such innovations can lower risks and provide solutions related to improved safety and security for both passengers and goods. In addition, there are emerging technologies improving the interoperability and multimodality of a transport network (see Chapter 5). There are new smart technologies providing access to information for all citizens, as well as technologies to reduce externalities such as transport noise. There are new smart security technologies and new fuels changing the historical reliance on natural resources (Auvinen *et al.*, 2012). These innovations together provide the possibility to reduce the drivers of climate change and to boost tele-information applications (Schade, 2016).

6.2.1.1 Road transport and fuel innovations

The notion of 'road transport innovation' includes a range of automotive solutions. The development of electric vehicles is strongly stimulated by the availability of long-range batteries, but brings risks leading to higher costs of electrification in the medium term (Contestabile and Alajaji Almubarak, 2017). Hybrid motors part powered by compressed air and hydrogen-fuelled vehicles are emerging innovative options, which may help to resolve environmental and energy-related issues. Hydrogen fuel cell vehicles especially have been the focus of increased attention in recent years. Hydrogen-fuel-cell-powered vehicles have a longer driving range without a long battery charging time and much lower on-vehicle emissions due to the direct conversion of energy in the fuel into electricity (Ehsani *et al.*, 2005). There are, however, two major barriers that must be overcome for the adoption of hydrogen fuel cell vehicles. The first is the cost of efficient and sustainable hydrogen production and delivery. The second is the requirement for new generations of hydrogen storage systems for vehicular and stationary applications (Edwards *et al.*, 2008). Separately, Fischer–Tropsch fuel is an alternative for heavy-duty vehicles. Synthesised from carbonaceous material, such as coal, natural gas or biomass, it has the potential to replace petroleum diesel fuel without any modification to a conventional diesel engine (Christensen *et al.*, 2010).

6.2.1.2 Road transport and safety

Public infrastructure constraints and the rising total vehicle distances travelled increase crash risk exposures. This prompts a requirement for new technologies that can seek to maximise traffic speeds while enhancing occupant safety. Autonomous vehicles have the potential to affordably reduce crashes by providing an alternative to drivers and by limiting speeds. In these ways, other road users can be protected. Self-driving technologies require additional equipment, higher roadway maintenance standards and special traffic signals; in addition, there is a need to adjust user attitudes to risk and technology acceptance (Kohl *et al.*, 2018).

This is not a simple task. Adjusting user attitudes involves addressing individual motivations and intentions: for instance, driving is not just a necessity, and for many may be an important reflection of their identity. Experiences with different technologies is another factor, as are other external variables. Proper information delivery by technology suppliers can reduce product users' perception of, for example, automated vehicles risk, as well as the actual risk of new road technology (Yong-Hui and Jing-Wen, 2009).

On the other hand, the autonomous vehicle provides the possibility of independent mobility for non-drivers, which can reduce stress and allows motorists to rest, play and work while travelling (Litman, 2018), but may also increase the number of vehicles on the road.

Platooning is one novel concept of intelligent transport dedicated to reducing the environmental impact of transportation, optimising fuel use and addressing safety issues, and which may also redress potential increases in vehicle numbers. In freight transportation, platooning involves connecting two or more heavy-duty vehicles in a convoy, using automated driving support systems. Boysen *et al.* (2017) indicated that the platoon requires a close synchronisation of schedules, individual truck routes and high levels of inter-communication standards. Although timing and movements are difficult to optimise, with each platoon manoeuvre needing to be performed in a compatible way, they may operate safely in any traffic situation. By using roads more effectively, intelligent platoon systems can deliver goods faster and reduce traffic jams (Kamali *et al.*, 2017; Maiti *et al.*, 2017). Similar platoon concepts may also be applied to cars.

6.2.1.3 Road transport and information and communication technology

The use of information and communication technology (ICT) has also attracted the interest of vehicle users and innovators in various transport sectors. For example, vehicle-to-vehicle (V2V) communication solutions permit safety-applications use data to be collected from nearby connected vehicles. As a consequence, we see one of the most powerful features of ICT; that is, the ability to collect information from across cars and transport infrastructures. This is complemented by the ability to store and process information, and to use and disseminate the information through information technology (IT) networks. Taken together, such developments have been used to develop road transport innovations (Dey *et al.*, 2016; Van Geenhuizen, 2009). Generally, V2V and vehicle-to-infrastructure (V2I) innovation and design guarantee the optimal use of available communication options and minimise crash risk and traffic problems, while optimising vehicle speeds, vehicle location management and traffic light operation (Merkisz, 2010).

At the other end of the technological innovation spectrum, and in the midst of organisational change and social transport innovations, a crucial role is played by car sharing and the ‘carpool’. A potentially large-scale mobility solution for cities seeking to reduce emissions, car sharing contributes to a reduction in the number of cars on the road, the density of peak-time public transport use and the related environmental burden. Car sharing is viewed as a simple and flexible system that does not depend on huge infrastructure investments, but that can be better facilitated with a range of of information and communications technologies.

6.2.1.4 Air, maritime and rail developments

Turning to air transport, it is fair to say that this sector mostly faces a very limited expansion of airport physical infrastructure, including at regional airports. There is, however, still plenty of scope for further improvements in air transport: manufacturers and research institutions are working on a large set of radical solutions, particularly for better fuel consumption, and hence improved emissions performance. Fuel savings are balanced against other crucial challenges regarding noise, minimising aircraft weight, system complexity and passenger acceptability (European Commission, 2011). Satellite technology, cybersecurity, directed energy, wireless intelligence and other smart city concepts all have huge influence on innovation in the air transport sector.

Historically, innovations in the air transport sector have varied in their success. Examples include rotorcraft, tiltrotor aircraft, airships (including dirigibles), eco-friendly planes and hydrogen cryoplane concepts. Innovations in infrastructure have included automated airports, 'highways in the sky' and off-shore air stations.

Increasingly, however, there is scope for smart technology applications in air transport: artificial intelligence can assist air traffic control; advanced computers can assist airline pilots – and perhaps could even replace them. An indicator of this might be unmanned aerial systems – 'drones' – or the intertwining of the internet of things, robotics and biometrics identification (Abeyratne, 2017). According to Salema *et al.* (2018), future aviation innovation could also include the blended wing body (conceived for very large aircraft, to achieve weight savings and higher aerodynamic efficiency) and truss-braced wings (aimed at reducing the induced drag of an increased payload). Improved technology in engine propulsive and aerodynamics design, and the use of lightweight materials to reduce aircraft weight and thermal efficiency could result in improved fuel savings and hence lower emissions (Teoh and Khoo, 2016). Minimising aircraft weight is a key consideration when designing an efficient air vehicle: one radical solution would replace windows with lightweight low-power-consumption screens (European Commission, 2011).

The marine transport sector supports more than 70% of global trade, so it is unsurprising that for many major economies, such as China and Japan, marine transportation is crucial. Among the innovative solutions in this sector, high-speed craft could be seen as a possible solution to current challenges. The fast maritime transport concept reduces the transit time, by increasing the overall transport process speed; reduces the associated freight costs, because the increased speed minimises the workshop costs; and optimises the order delivery process with secure and on-time delivery of shipments, thus decreasing the logistical chain global costs (Fernando and Chukai, 2018; Martínez de Osés and Castells, 2005).

Other elements of discussion on marine transport, with particular emphasis on intercontinental transport, are containerisation, container-ship size and speed, and the Automatic Identification System (AIS). Containerisation has become a fundamental constituent of large-scale marine transport (Leea and Song, 2017). An analysis of global trends shows that the ocean carriers providing maritime transport services to shippers should consider that fuel consumption by a container ship is mostly a function of ship size and cruising speed (Notteboom and Carriou, 2009). Maritime emissions play an important role in anthropogenic city emissions for those cities with busy ports. Ship engines pollute the marine environment. Ship owners, using the AIS, are able to provide reliable speed data for maritime emissions management (Yau *et al.*, 2012). It is possible to profile the general vessel speed and to thereby estimate emissions. In addition to improving the operational efficiency of conventional marine fuel consumption and emissions, there is also growing interest in the possible adoption of more environmentally friendly hydrogen fuel cell systems (Kauf, 2012).

The operational efficiency of marine vessels and fleets refers also to the securing of safe transportation. In recent years, the digitisation and automation of ships has been accelerating. The shipbuilding and shipping industries integrate ICT technology by using the internet of things, cloud computing, big data and mobile security (Im *et al.*, 2018). In both cases the infrastructure on the ships and on shore (the e-navigation server, vessel traffic service (VTS) stations and the land-based AIS) are connected via the internet (Filipkowski, 2013). A ‘smart ship’ is equipped with numerous ICT technologies monitoring the ship and its functions, such as the navigation systems, safety systems, machinery, automation, cargo environment and the weather conditions in the area of operation (Kobyliński, 2017). Remote and autonomous ships have the potential to reduce human-based errors and, at the same time, may modify some existing risks (e.g. in communication and connectivity). Integrating intelligent IT, smart ships (remote controlled or fully autonomous) and shore architecture have the potential to be increasingly economical, safe and efficient. However, questions of liability for smart ships are a subject for national discussion and research (Rolls-Royce, 2016).

Technological innovations have also improved the process of unloading and loading a ship. Automated container handling (terminal) technology and automated material handling equipment are high-value investments developed during the past decades. Automated cranes and vehicles perform their tasks without human intervention, and automated control technology has become standard for controlling material-handling equipment, improving efficiency, reducing costs and increasing capacity (Vis and Harika, 2004; Zhena *et al.*, 2018).

In the rail supply industry, competition is high despite the rather limited number of players and the relatively small market size in comparison with road transport modes. Railway vehicles, signalling and control systems have typical lifespans of around 30–35 years. These factors ensure that the market for new locomotives and rolling stock is small, and the development of major technological changes is very difficult. Innovation is further hampered by the particularly high innovation costs in this industry (Wiesenthal *et al.*, 2015).

Rail innovation initially evolved from the need to increase safety for railways and to meet an expected shift towards rail (Hansen *et al.*, 2016)

> [Its] potential to support a low carbon economy, higher capacity, reliability, energy efficiency, security standards, as well as lower operating costs and additional services for customers are all beneficial. Automation also provides for higher safety standards. Single processes have been automated, such as rail signaling and logistic scheduling. For public transport applications, such as metro trains, whole systems have become fully automated based on automatic train protection, operation and train controls (including: safety functions as red signal overrunning, speed limits and collision prevention systems that are fully automated).

One innovative project in the rail sector focuses on the development of a real-time intelligent traffic management system with in-cab signalling to reduce headways and to allow decreased traction energy consumption. The precise location, speed, breaking and load of every train on the network is optimised to improve operational performance and to keep the passengers informed about service performance. Other innovations include energy recovery systems in rolling stock and alternative fuels, allowing train operators to achieve lower operating costs (TSLG, 2012). Other rail-related innovative projects include double-deck high-speed trains, tram–train synergies, train scheduling optimisation and hybrid trains.

6.2.1.5 Infrastructure and urban transport innovations

Transport development is also influenced by infrastructure innovations, including alternative materials in road construction, such as recycled materials, skid-resistant road surfaces, surface de-icing improvements and new noise-absorbing road surfaces (Burnewicz, 2010). Such infrastructure innovations may be particularly advantageous to citizens in metropolitan areas, which may sustain significant negative effects from conventional transport emissions, congestion, noise and so on. These areas may also benefit from more responsive transport innovations: low-emission zones, urban lift sharing and public transport services (International Transport Forum, 2016).

Urban transport innovation also includes optimally planning, managing and controlling a metropolitan area. Organisational innovations address urban problems such as driving directions, traffic signals and restrictions, and speed limits, with the objective of optimising the travel choice and minimising the total time spent by all travellers (Amaral and Aghezzaf, 2015). Providing drivers with real-time information, such as transit routes and schedules, navigation directions, information about delays due to congestion, accidents, weather conditions and road repair work, improves time spent driving, and thereby reduces fuel consumption and emissions. Telematic applications (Schade, 2016) can help with traffic control, and adjust urban infrastructure to address increased vehicle traffic and demand for parking spaces (Parkitny, 2016). Today, if someone wants to drive into a city, or use public transport, they simply use mobile communication technologies both for payment and real-time information (Sokołowski, 2016).

6.3. ITSs: characteristics and reasons for implementation

An ITS aims to apply intelligent information and communication technologies in order to improve transport services and traffic management. Key objectives of ITS development include achieving sustainable mobility through intermodality, improving access to all places within the city, mitigating the impact of road traffic, increasing a modal share of transport and making better use of existing networks. This impacts on both traffic information control (vehicle control, traffic control, pre-trip and on-trip information) and management services (passenger services, emergency service system, mayday services and demand management) (McDonald *et al.*, 2006). Innovations in urban or metropolitan transport focus on traffic control, air pollution, crime control, efficient navigation, congestion control, resource management and disaster management (Zear *et al.*, 2016). Fully integrated intelligent transportation systems, such as V2I and V2V integration, enable communication among assets in the transportation system. ITSs can increase safety by providing local danger warnings, real-time traffic and travel information and speed alerts, as well as other enhancements, such as eco-driving coaching and improved operational performance, particularly by reducing congestion, thereby improving mobility and travel convenience. Connected to these improvements are contingent productivity, economic and employment boosts (Ezell, 2010).

The EU is committed to the increasing use of ITSs, as evidenced in its inclusion in a number of important documents, notably the 2018 European White Paper on transport (EU, 2018) and European Directive 2010/40/EU (European Council, 2010). A touted example of ITSs is the eCall system that will ultimately be available in the entire EU (European Commission, 2019a). eCall will automatically phone for help in the case of a traffic collision or accident, using satellite communications systems for location purposes.

6.3.1 Transport in Europe and the case of Łódź

The number of private cars in the EU in comparison with the number of its inhabitants is high – in 2016 it ranged from 662 (Luxembourg) to 279 (Romania) cars per 1000 inhabitants. Poland is near the top in this ranking, occupying the sixth position (out of 28), with 571 cars per 1000 inhabitants. Excepting Luxembourg, this number is only higher in the following countries: Italy (625), Malta (615), Finland (606) and Cyprus (595). Poland is followed by such countries as Austria (550), Germany (546), Belgium (503), the Netherlands (481) and France (479) (Eurostat, 2016). Such a high number of private cars in the EU requires efficient traffic control. Apart from traditional solutions, including traffic control by the police or traffic wardens, modern technological solutions can be employed. Today, ITSs are most frequently used in cities where traffic control is difficult. In Poland, a further pressing reason for the introduction of ITSs is to improve road safety, as Poland has one of the highest number of accidents of all the EU member states (European Road Safety Observatory, 2017).

There are also many other reasons why an ITS is implemented, such as the potential to use it for balancing economic development on a regional or national level (Kozłowski and Marczak, 2014). Thus, fast-developing countries should adopt these solutions on a large scale. When choosing the ITS type and functionalities for individual countries, or even at the city level, it is necessary to take into consideration the specific nature of drivers' behaviour (driving etiquette) and the characteristics of the transport system, reflected in the infrastructure.

Unlike many of its western EU neighbours, Poland has virtually no underground rail systems (exceptions are two underground lines in Warsaw). This has to be considered both when selecting an ITS and later, when it is installed and put into operation. Łódź is the third largest city in Poland, located nearly in the centre of the country. Łódź is where most of the country's important expressways and motorways intersect (Falqueto *et al.*, 2007), and, after the cross-city railway tunnel is completed, it will also become the most significant railway junction in Poland. Such a location needs to be taken into consideration in the ITS, which should deal with both incoming and outgoing streams of cars, and with passenger railways as well as interchange within the city to mass transport or taxis.

In the case of the transport system in Łódź, the entirety of its traffic is concentrated at ground level – remember that, unlike Munich or London, the city has no underground rail services. Another characteristic of the city is the very low (for an urban area with more than 1.1 million inhabitants) number of engineering structures for streamlining traffic, such as tunnels or flyovers. For this reason, there is very high congestion. According to TomTom (2019), Łódź is the most congested city not only in Poland but second only to Dublin in the entire EU in 2018. It also occupies the 15th position in the world. From TomTom's statistics, Łódź has a congestion level (the extra travel time needed compared with uncongested conditions) of 44%.

Solutions to the problem are mostly concerned with the development of transport infrastructures: the construction of numerous flyovers to streamline traffic; park-and-ride facilities to reduce city centre congestion; two cross-city railway tunnels for conventional and high-speed rail; and surface street infrastructure supporting mass and individual green transport, such as the use of bicycles. While the contextual complexity and scale of the congestion problems in the city mean that an ITS can only be an auxiliary tool to material infrastructural change, it is nevertheless important to consider.

6.3.2 An innovative ITS in Łódź

The purpose of the ITS in Łódź is to process and share information that can prevent potential crashes, keep traffic moving, and decrease the negative environmental impacts of the transportation sector in the city.

The construction of the Łódź ITS was carried out on the occasion of the expansion of the regional tram, and was launched in 2015, and then was slightly extended. There was a series of aims to the project, including social, economic and environmental, such as better support of public transport (700 public transport vehicles had 'smart' terminals installed, allowing them to connect to the ITS), shortening of travel times, reducing the number of congestion events and traffic jams, improving road safety, and lowering noise and pollution levels. Operation of the ITS in Łódź should lead to a reduction of public transport travel times while increasing the comfort of travel.

During its construction, 555 km of electrical cables were laid and 40 km of cable conduits were built. At present, the ITS is the largest operating system of its kind in Poland, consisting of

- more than 250 junctions with inductive loops detecting four- and two-wheeled vehicles
- inductive loops installed at pedestrian crossings and bike lanes, detecting pedestrians and cyclists
- 89 CCTV cameras
- 9 variable-message signs (mobile information boards for drivers)
- 226 automatic number-plate-recognition cameras

- 150 passenger information boards at 144 bus and tram stops
- a system of cameras detecting dangerous situations in the city's road tunnel
- other traffic detectors and engineering devices (Kozłowski *et al.*, 2016).

Some of the key elements of the ITS are sensors collecting road traffic data. In order to identify individual traffic participants, the following dedicated devices are used: microwave detectors in the case of cyclists, sensor keys in the case of pedestrians, and inductive detectors and video detectors in the case of vehicles.

The system elements dedicated to supporting public transport include passenger information boards located at bus and tram stops, displaying real-time information about the arrival times of public transport vehicles.

A web-based interactive online map for drivers and public transport passengers includes

- meteorological information
- a list of works conducted that hinder communication in the city
- a tool allowing a route to be shown after providing the start and end points
- a map visualising the communications situation in the city (Figure 6.1).

The whole system is managed by the Traffic Control Centre. This has more than 30 large monitors on which operators, divided into two teams, the Aereal Traffic Control System

Figure 6.1 Map from the ITS Łódź portal (ITS Łódź, 2019)

Team and the Visual Surveillance System Team, watch the traffic in the streets. The system recognises such elements as number plates and vehicle characteristics, which enables categorisation of each vehicle in accordance with the vehicle-type classification. It stores camera footage and infrared images. Along the east–west route, it automatically identifies adverse events in the road tunnel, which is particularly important, as this is a crucial point in the city's infrastructure.

6.4. Opportunities provided by the ITS in Łódź and further developments

After the ITS was launched in Łódź, a series of measurements were taken in order to assess the effects of its implementation. The measurements were carried out by the Road and Transport Office in Łódź and the system provider Sprint. The results demonstrated that the average travel time was shorter by 9.6% in the case of public transport and by as much as 29.5% in the case of private cars. The study also showed a large decrease in the number of traffic jams (Kamiński and Oskarbski, 2017). Owing to the ITS, the city's transport infrastructure capacity was improved without extensive and costly structural changes, which, due to existing buildings, would also be impractical in many places.

As a result of the system implementation, safety also increased considerably. In the case of an emergency, the Traffic Control Centre deploys help as quickly as possible, and then works to relieve traffic jams that form in such situations. Some incidents are also avoided, not least perhaps because drivers are aware of the fact that their behaviour is recorded, so any traffic violation may be quickly proven. The police, municipal police and other services have the means to detect offences and identify offenders.

Drivers in Łódź tend to speed. The reason for this is that they spend a lot of time in traffic jams. Consequently, when they see an uncongested road, they try to make up for lost time by exceeding the speed limit. This is a very serious barrier to using the so-called 'green wave' of managed and co-ordinated traffic flow, which in other cases could reduce travel irregularities and congestion. Therefore, after analyses of the data from the Łódź ITS, the green-wave approach has been adopted only sparingly.

Another effect of the high congestion in Łódź is that cars stop in the middle of junctions during morning and afternoon rush hours, blocking them, which considerably limits the traffic capacity, additionally increasing traffic jams. The ITS offers a solution: it constantly analyses the situation at junctions and 'allows' only as many cars as will pass through a junction without blocking it.

Interestingly, and despite significant successes, there is dissatisfaction with the functioning of the ITS system. No doubt in part due to the growing number of private vehicles (a trend set to continue), the system is not able to fully control high-volume traffic. The main reason is the underdevelopment of the transport infrastructure in Łódź in the form of

- information provision
- tunnel infrastructure
- roads planned, but not yet built.

New sections of the S14 expressway and roads leading to it need to be completed. There are also plans to extend the interchange at the largest railway station in Poland – Łódź Fabryczna Rail Station – and to build an interchange at the international airport in Łódź. A further factor for Łódź is the planned urban revitalisation that will affect a considerable part of the city. This will result in various traffic difficulties and the necessity to reorganise the traffic a number of times during the redevelopment process, as well as permanent changes in traffic organisation as a result of modernisation and alteration of some streets located in the area.

6.5. Recommendations

The situation in Łódź presents a huge challenge for those in charge of the city traffic. In order properly to manage the city's transport, it will be necessary to find both substantive and technological support.

Substantive support should be based on similar solutions to those that have already been adopted. One of the most popular proposed types of support is the creation of a dedicated team for traffic analyses that puts forward recommendations on how to manage the traffic so that it is adjusted to the specific character of the city. The team would support the services already operating in the Traffic Control Centre that directly manages the Łodź ITS. It should consist of academics and independent analysts, which means it should be an 'external' unit. Wider analysis of the achievements and shortcomings of the Łodź ITS could help ensure that the affordances of ITS – including opportunities for refinement, expansion and evolution – are identified and achieved.

In technological terms, the next stage of development of the Łódź ITS will be an evolution towards a cooperative ITS (C-ITS). Systems of this kind (as described in Section 6.2.1) allow technologies installed in vehicles to communicate with

- elements of the road infrastructure (V2I)
- technologies installed in other vehicles (V2V)
- other elements of the transport system.

The EU places great emphasis on the development of C-ITSs. Its strategic programmes include the construction of such systems as the basis for transport logistics management. The EU is committed to C-ITSs because it believes such systems lead to

- increased safety
- improved traffic flow
- shorter travel times
- better use of the transport infrastructure.

Such outcomes would lead also to improved metropolitan environments, with reduced fuel consumption and emissions, as well as increased social goods, such as travel time efficiencies and fewer accidents. By developing ITSs towards C-ITSs, cities will prepare for new challenges and improve the standard of living of their citizens. In short, C-ITSs are essential to the future of technologically modern and environmentally smart cities.

In the more distant future, further and more extensive changes will occur on account of autonomous vehicles joining the traffic. Autonomous vehicles will provide people with a number of new opportunities as well as new transport and transport-related services. C-ITSs will be increasingly helpful in managing them.

The evolution of the Łódź traffic control system towards a C-ITS will require considerable financial outlay. However, as C-ITSs are a strategic direction indicated by a number of important EU policy documents (e.g. see European Commission, 2019b), it should be possible to obtain funds from the EU.

In closing, the mobility challenge can be met by creative solutions that, as presented in the present chapter, improve safety, speed and coverage, using emerging technologies for social and environmental benefit and improving quality of life.

REFERENCES

Abeyratne R (2017) *Megatrends and Air Transport Legal, Ethical and Economic Issues*. Springer, Berlin, Germany.

Amaral RR and Aghezzaf ElH (2015) City logistics and traffic management: modelling the inner and outer urban transport flows in a two-tiered system. *Transportation Research Procedia* **6**: 297–312.

Auvinen H, Tuominen A and Ahlqvist T (2012) Towards long-term foresight for transport: envisioning the Finnish transport system in 2100. *Foresight* **14(3)**: 191–206.

Boysen N, Briskorn D and Schwerdfeger S (2017) The identical-path truck platooning problem. *Transportation Research, Part B* **109**: 26–39.

Burnewicz J (2010) Perspektywa innowacyjna transportu i logistyki. *Ekonomiczne Problemy Usług* **59**: 51–63.

Christensen J, Glaeser KP, Shelton T, Moore B and Aarts L (2010) *Innovation in Truck Technologies*. OECD Publishing, Paris, France, OECD/ITF Joint Transport Research Centre Discussion Papers, No. 2010/10. https://www.oecd-ilibrary.org/transport/innovation-in-truck-technologies_5kmjp66069s4-en (accessed 20/06/2019).

Contestabile M and Alajaji Almubarak B (2017) Will current electric vehicle policy lead to cost-effective electrification of passenger car transport? *Energy Policy* **110**: 20–30.

Dey KCh, Rayamajhi A, Chowdhury M, Bhavsar P and Martin J (2016) Vehicle-to-vehicle (V2V) and vehicle-to-infrastructure (V2I) communication in a heterogeneous wireless network – performance evaluation. *Transportation Research, Part C* **68**: 168–184.

Dobranskyte-Niskota A, Perujo A, Jesinghaus J and Jensen P (2009) *Indicators to Assess Sustainability of Transport Activities*. Joint Research Centre Institute for Environment and Sustainability, European Commission, Luxemburg.

Edwards PP, Kuznetsov VL, David WIF and Brandon NP (2008) Hydrogen and fuel cells: towards a sustainable energy future. *Energy Policy* **36**: 4356–4362.

Ehsani M, Gao Y, Gay SE and Emadi A (2005) *Modern Electric, Hybrid Electric, and Fuel Cell Vehicles. Fundamentals, Theory, and Design*. CRC Press, London, UK.

EMTA (European Metropolitan Transport Authorities) (2017) *2015 Barometer*, 11th edn. EMTA, Paris, France. https://www.emta.com/spip.php?article267 (accessed 20/06/2019).

EU (European Union) (2018) *White Paper on Transport. Roadmap to a Single European Transport Area – Towards a Competitive and Resource-efficient Transport System*. EU,

Luxembourg. https://ec.europa.eu/transport/sites/transport/files/themes/strategies/doc/2011_white_paper/white-paper-illustrated-brochure_en.pdf (accessed 20/06/2019).

European Commission (2011) *Flightpath 2050: Europe's Vision for Aviation. Report of the High Level Group on Aviation Research.* European Union, Luxemburg. https://ec.europa.eu/transport/sites/transport/files/modes/air/doc/flightpath2050.pdf (accessed 20/06/2019).

European Commission (2019a) eCall. https://ec.europa.eu/transport/road_safety/specialist/knowledge/esave/esafety_measures_unknown_safety_effects/ecall_en (accessed 20/06/2019).

European Commission (2019b) Cooperative, connected and automated mobility (CCAM). https://ec.europa.eu/transport/themes/its/c-its_en (accessed 20/06/2019).

European Council (2010) Directive 2010/40/EU of the European Parliament and of the Council of 7 July 2010 on the framework for the deployment of Intelligent Transport Systems in the field of road transport and for interfaces with other modes of transport. https://eur-lex.europa.eu/legal-content/EN/ALL/?uri=CELEX%3A32010L0040 (accessed 20/06/2019).

European Road Safety Observatory (2017) *Traffic Safety. Basic Facts 2017.* https://ec.europa.eu/transport/road_safety/sites/roadsafety/files/pdf/statistics/dacota/bfs2017_main_figures.pdf (accessed 20/06/2019).

Eurostat (2016) Number of passenger cars per 1000 inhabitants, 2016. https://ec.europa.eu/eurostat/statistics-explained/index.php/Stock_of_vehicles_at_regional_level (accessed 20/06/2019).

Ezell S (2010) Explaining international IT application leadership: intelligent transportation systems. ITIF. https://itif.org/publications/2010/01/09/explaining-international-it-application-leadership-intelligent (accessed 20/06/2019).

Falqueto T, Nuttall WJ and Trzmielak DN (2007) An assessment of logistic system opportunities for Łódź (Lessons learned from the Zaragoza experience). *Marketing technologiczny i marketing terytorialny* **235**: 324-346.

Fernando Y and Chukai Ch (2018) Value co-creation, goods and service tax (GST) impacts on sustainable logistic performance. *Research in Transportation Business & Management* **28**: 92–102.

Filipkowski D (2013) Data transmission system architecture for e-navigation. In: *Activities of Transport Telematics. 13th International Conference on Transport Systems Telematics, TST 2013* (Mikulski J (ed.)). Springer, Berlin, Germany, Communications in Computer and Information Science, vol. 395.

Hansen Ch, Daim T, Ernst H and Herstatt C (2016) The future of rail automation: a scenario-based technology roadmap for the rail automation market. *Technological Forecasting & Social Change* **110**: 96–212.

Im I, Shin D and Jeong J (2018) Components for smart autonomous ship architecture based on intelligent information technology. *Procedia Computer Science* **134**: 91–98.

International Transport Forum (2016) *Zero Road Deaths and Serious Injuries: Leading a Paradigm Shift to a Safe System.* Organisation for Economic Co-operation and Development, Paris, France. http://www.carsp.ca/carsp-acpser/wp-content/uploads/2017/01/OECD-Safe-Systems.pdf

Irwin DA (2000) *Free Trade under Fire.* Princeton University Press, Princeton, NJ, USA.

ITS Łódź (2019) http://www.its.lodz.pl (accessed 20/06/2019).

Kamali M, Dennis LA, McAree O, Fisher M and Veres SM (2017) Formal verification of autonomous vehicle platooning. *Science of Computer Programming* **148**: 88–106.

Kamiński T and Oskarbski J (2017) Wdrażanie systemów ITS na przykładzie miasta Łodzi i Bydgoszczy. *Transport samochodowy* **1**: 28.

Kauf S (2012) Innowacyjne rozwiązania w logistyce transportu, Przedsiębiorczość i Zarządzanie. *Nowe kierunki w zarzadzaniu logistycznym, Przedsiębiorczość i Zarządzanie* **18(12-1)**: 171–188.

Kobyliński L (2017) Smart ships – autonomous or remote controlled? *Scientific Journals of the Maritime University of Szczecin* **53(125)**: 28–34.

Kohl Ch, Knigge M, Baader G, Böhm M and Krcmar H (2018) Anticipating acceptance of emerging technologies using Twitter: the case of self-driving cars. *Journal of Business Economics* **88(5)**: 617–642.

Kozłowski R and Marczak M (2014) Budowa inteligentnych systemów transportowych jako szansa dla zrównoważonego rozwoju regionów. *Ekonomia i Zarządzanie* **6(2–6)**: 37–40.

Kozłowski R, Jabłoński J and Palczewska A (2016) The launching phase analysis of the implementation of Polish largest intelligent transportation system. *Archives of Transport System Telematics* **9(2)**: 27.

Leea ChY and Song D (2017) Ocean container transport in global supply chains: overview and research opportunities. *Transportation Research, Part B* **95**: 442–474.

Litman T (2018) *Autonomous Vehicle Implementation Predictions. Implications for Transport Planning*. Victoria Transport Policy Institute, Victoria, Australia. https://www.vtpi.org/avip.pdf (accessed 20/06/2019).

McDonald M, Keller H, Klijnhout J *et al.* (2006) *Intelligent Transport Systems in Europe: Opportunities for Future Research*. World Scientific, Singapore.

Maiti S, Winter S and Kulik L (2017) A conceptualization of vehicle platoons and platoon operations. *Transportation Research, Part C* **80**: 1–19.

Martínez de Osés FX and Castells M (2005) High speed craft viability analysis. *Journal of Maritime Research* **3(2)**: 59–76.

Merkisz J (2010) Inteligentne rozwiązania systemów informacyjnych w pojazdach osobowych i ciężarowych. *Problemy transportu i logistyki. Innowacje w transporcie organizacja i zarządzanie* **602(12)**: 165–180.

Notteboom T and Carriou P (2009) Fuel surcharge practices of container shipping lines: Is it about cost recovery or revenue making? *Proceedings of the 2009 International Association of Maritime Economists (IAME) Conference, Copenhagen, Denmark.*

Parkitny W (2016) Problemy stojące przed logistyką miejską w zakresie obsługi parkingowej pojazdów. *Przedsiębiorczość i Zarządzanie* **18(12-1)**: 309–320.

Rolls-Royce (2016) *Autonomous Ships. The Next Step*. Rolls-Royce, London, UK. https://www.rolls-royce.com/~/media/Files/R/Rolls-Royce/documents/customers/marine/ship-intel/rr-ship-intel-aawa-8pg.pdf (accessed 20/06/2019).

Salema KA, Binanteb V, Cipollab V and Maganzic M (2018) PARSIFAL Project: a breakthrough innovation in air transport. *Aerotecnica* **97(1)**: 40–46.

Schade W (2016) Comparison of innovation systems of different transport modes and the need for public intervention. *Transportation Research Procedia* **14**: 4105–4112.

Sokołowski W (2012) Analysis of elektronic and tool system in selected EU countries. *Przedsiębiorczość i Zarządzanie* **18(12-1)**: 379–399.

Teoh LE and Khoo HL (2016) Green air transport system: an overview of issues, strategies and challenges. *KSCE Journal of Civil Engineering* **20(3)**: 1040–1052.

TomTom (2019) Traffic index 2018. https://www.tomtom.com/en_gb/traffic-index/ranking/ (accessed 20/06/2019).

TSLG (Technical Strategy Leadership Group) (2012) *The Future Rail Way. The Industry's Rail Technical Strategy 2012.* Rail Transport Service, London, UK. https://www.rssb.co.uk/Library/Future%20Railway/innovation-in-rail-rail-technical-strategy-2012.pdf (accessed 20/06/2019).

Van Geenhuizen M (2009) Reflections on ICT in world city development and development of remote regions. In: *Value-Added Partnering and Innovation in a Changing World* (Van Geenhuizen M, Trzmielak DM, Gibson DV and Urbaniak M (eds)). Purdue University Press, West Lafayette, IN, USA, pp. 236–251.

Van Geenhuizen M and Schoonman J (2010) Renewable energy technologies and potential for adoption. In: *Energy and Innovation Structural Change and Policy Implications* (Van Geenhuizen M, Nuttall WJ, Gibson DV and Oftedal EM (eds)). Purdue University Press, Lafayette, IN, USA, pp. 17–36.

Vis IFA and Harika I (2004) Comparison of vehicle types at an automated container terminal. *OR Spectrum* **26**: 117–143.

Wiesenthal T, Condeço-Melhorado A and Leduc G (2015) Innovation in the European transport sector: a review. *Transport Policy* **45**: 86–93.

Yau PS, Lee S and Ho KF (2012) Speed profiles for improvement of maritime emission estimation. *Environmental Engineering Science* **29(12)**: 1076–1084.

Yong-Hui L and Jing-Wen H (2009) Applying theory of perceived risk and technology acceptance model in the online shopping channel. *International Journal of Economics and Management Engineering* **3(5)**: 505–511.

Zear A, Singh PK and Singh Y (2016) Intelligent transport system: a progressive review. *Indian Journal of Science and Technology* **9(32)**: 1–8.

Zhena L, Hub H, Wanga W, Shic X and Ma Ch (2018) Cranes scheduling in frame bridges based automated container terminals. *Transportation Research, Part C* **97**: 369–384.

Part 3. Energy

ice Publishing

Nuttall, William J, Gibson, David V, Trzmielak, Dariusz and Ibarra-Yunez, Alejandro
ISBN 978-0-7277-6425-6
https://doi.org/10.1680/emsc.64256.117

Chapter 7
The effects of renewable energy policies in electricity markets: global perspectives

Alejandro Ibarra-Yunez
Tecnológico de Monterrey, Mexico

7.1. Introduction

In recent years, many governments have increased the competences and administrative power of energy regulators and climate change agencies, resulting in a somewhat asymmetric and uncoordinated effort by the two institutions and their objectives. On the one hand, energy regulators are in charge of market activities in electricity wholesale markets, increasingly integrating renewable sources of energy to shift the energy mix. In addition they guarantee *ex ante* (or sectoral) and *ex post* (or competition policy) regulation to monopoly segments of the market at the same time that they ensure market-driven investments (Griffin and Puller, 2005; Hogan, 2002; Li *et al.*, 2015; Pollitt, 2008, 2012). On the other hand, renewable energy targets and climate change policies are set by agencies with their own objectives, mostly regarding greenhouse gas emissions reduction and carbon dioxide footprint abatement via recycling, mitigation and sequestration of pollutants, in urban (mainly) and rural (less so) areas. These two sets of policies could move on parallel paths such that emission reduction is not excessively costly, while the deepening of the energy market creates incentives for cost-efficient dissemination for investments that integrate renewable energies (EERE, 2012; Kessides and Wade, 2011; Li *et al.*, 2015; Sivaraman and Moore, 2012). Moreover, both sets of policies could be integrated into urban planning and smart city objectives. The challenge, then, is to ensure that targets do not contradict each other and that the respective agencies face convergent and complementary objectives (Coen and Thatcher, 2008; Egerer *et al.*, 2015; Executive Office of the President of the United States, 2016; Rouhani *et al.*, 2016).

However, there is the risk that the two sets of policies may not incentivise investment and production efforts in a new wave of clean and renewable energy production; so, the two sets could even provoke a setback in investment and production efforts, due to asymmetrical objectives or policy instruments. The question of a set of superimposed regulatory agencies is rather new in the literature but has a significant impact on the coordination of delegated regulators to deepen markets for clean fossil sources of electricity generation and increase renewable sources to change the energy mix overall, and in visionary objectives in smart city agendas (EERE, 2012; Rouhani *et al.*, 2016; Vidal-Amaro *et al.*, 2015).

The present research uses a unique data set applied to a group of 34 countries (18 emerging and 16 OECD developed economies) between 2004 and 2015, to quantify the effectiveness of governments and regulatory quality indices on renewable production efforts and their dynamics. The indices are from the Worldwide Governance Indicators (World Bank, 2016).

> [The Government Effectiveness index] reflects perceptions of the quality of public services, the quality of the civil service and the degree of its independence from political pressures, the quality of policy formulation and implementation, and the credibility of the government's commitment to such policies. The Regulatory Quality index reflects perceptions of the ability of the government to formulate and implement sound policies and regulations that permit and promote private sector development.

The two indices are standardised around a mean for cross-country comparisons.

The results when applied to entire countries could have implications for specific urban areas and markets, such as distributed generation, smart grid development, and 'prosumers' of energy in an autonomous manner.

Additionally, the panel data analysis integrates data from a measure of market power by incumbent utilities and a move towards more market competition as facilitating new investment and production efforts in renewable energies, separated by technology (hydroelectric, wind, and solar photovoltaic (PV) generation), by country and by country size.

The dynamics of an accelerated renewable output, mainly for wind and solar PV technologies, is integrated into the analysis to separate endogenous increases in production from those incentivised by the sets of regulatory quality and effectiveness of government indices, as conditions for further smart city strategies. In this respect, the present analysis is new and aims to contribute to the understanding of policy architecture to trigger strong economic efforts by all actors towards a renewable energy mix. From the literature review, no previous study has used government effectiveness and regulatory quality indices to partially explain investment in and production of renewable energies.

The results show that government effectiveness and regulatory quality are indeed significant in explaining the uptake of wind and solar renewable generation, additionally from their own explosive renewable dynamic uptake. The author has found that dynamic solar PV output growth is relatively more affected by specific regulatory quality measures. Overall, reduced market power by incumbents at the moment of reforms also significantly explains the uptake of renewables output. However, the clarity of governance quality is much more evident in developed economies, while evidence is muted in developing countries, where, over recent years, some governments have seen regulatory quality and government effectiveness indices diminish. That sets policy challenges for renewables, in conjunction with financing smart city objectives and planning.

This chapter is organised as follows. After a review of the relevant literature in the next section, Section 7.3 describes and analyses the time path of the launch of renewable energies (wind, solar PV, hydroelectric generation) in developed and developing countries, showing the explosive growth of wind and, later, solar sources of energy, but lesser growth for more traditional hydroelectric generation; Section 7.4 describes the model used for the dynamic and explosive production of renewable energy, for both wind and solar PV with their determinants of the quality of regulations and government efficiency. A panel data model was estimated using various specifications to account for the quality indices, the market power of incumbent utilities versus competition, and control variables for country size, country type, and break year of the beginning of explosive growth in renewable production. The results are analysed and compared with few technicalities for a general, non-mathematical audience. Section 7.5 extends the analysis and implications of the indirect effect of regulatory quality on renewable production and renewables' dynamics, and implications are discussed for smart cities' planning agendas. Section 7.6 concludes.

7.2. Overall framework for analysis

Renewable energy, its trends and dynamics of investment, and the launch of so-called clean projects (mostly hydroelectric, wind and solar sources) have been widely studied in economics and give credence to the launch of renewable energy beginning in the 1990s, mainly in developed economies, divided by world regions. Notably, economics papers on the explosive growth of wind and solar energies, while showing other less dramatic changes in hydroelectric and nuclear power, such as by Aguirre and Ibikunle (2014), have used a modified trend (autoregressive) model to check for renewable growth in a sample of country cases for a long period: 1990–2010. Ata (2016) used a set of renewable policy instruments (such as feed-in tariffs) to incentivise the high growth of renewables, mostly in the EU, but with lessons for the USA, while Basher *et al.* (2015) analysed unit roots from the trend of the uptake of renewable energy in a group of OECD countries. All analyses are recent and aim to understand the somewhat endogenous growth in renewables, independent from policy or market characteristics. For its part, the literature on smart cities has concentrated on agents empowered to produce changes in planning, transport systems, the role of distributed energy as models of sustainability, big data as a critical resource, the governance of smart cities and community involvement, and even shifts in city amenities for cities of the future (Albino *et al.*, 2015; Chourabi *et al.*, 2012).

The present chapter focuses on excess government influence, which has been investigated under various theoretical frameworks. To begin with, the theory of multi-principals has been studied since the mid-1980s as a source of bureaucratic regulations where a set of agencies with asymmetric endowments and administrative plus instrumentation capacities could give rise to market and regulatory failure since market power does not diminish after enacting regulations. Under such a framework, the incumbent utility could under-invest or deny access to basic infrastructure, or a regulator faces diminishing returns after some time of regulating for the market, such as reported by Estache and Martimort (1998) and Spiller (1990). The causes of government regulatory imperfections have also been studied under commitment and no commitment of government framers towards a certain market (Alchian and Demsetz, 1972; Laffont and Tirole 1991).

For electricity markets, Hogan *et al.* (2010) have addressed the optimal combination of regulated and market mechanisms for transmission expansion. The wave of replicated deregulation efforts has been studied by Jordana and Levi-Faur (2005), Majone (1997, 2005) and Pollack (2003), while the dynamics of electricity market unbundling have been addressed by Adelman and Spence (2015), Griffin and Puller (2005) (for Europe), Harris (2006), Jacobsson and Lauber (2006), and Pollitt (2008, 2012).

One can conceptualise networks of regulators following 'superior' laws, such as ones arising from the European Commission, or from Mexico's Executive and its Secretary of Energy when proposing constitutional changes in 2013 for energy reform and then 'delegating' jurisdictional endowments to the regulators in charge of applying the 'secondary' or lower level laws. These can be viewed as a pyramid of higher regulations that spread down to laws, directives, and the specific manuals of operation in a 'new market', part of which keeps regulated segments (e.g. electricity basic services subject to price controls, transmission and part of the distribution grids). Emerging economies, such as China, India, Brazil or Costa Rica, have also defined areas of renewable promotion, specifically treated and separated from energy development incentives. Additionally, under institutional economics some have followed a principal-agent methodology, with cooperative versus non-cooperative agreements in games of commitment with time asymmetries as reported by Coen and Thatcher (2008). The main thrust there is on how central versus decentralised government coexists, and helps to develop energy self-sufficiency in countries, states, localities and urban communities.

Turning now to quantitative analyses of regulatory impacts, the main contributions are summarised by, for example, the following: Bunn and Muñoz (2015) on intermittent modelling; Bushnell and Stoft (1997) and Borenstein *et al.* (2002) on congestion and balancing mechanisms in California; Egerer *et al.* (2015) on network expansion and regulation; Gabriel *et al.* (2013) on complementary modelling and optimisation; Pollitt *et al.* (2007) on regulation and vertical unbundling; and Wolak (2015) on simulation approaches to congestion. Many other analyses concentrate on the simulation of renewable energy trends, and optimisation, while fewer focus on econometric modelling (Egerer *et al.*, 2015; Li *et al.*, 2015; Tomain, 2016).

7.3. Growth trends in renewable energies

The stage of development in general determines for how long countries have embarked on renewable deployment, as an indication of market maturity. It is clear that one cannot separate the origin and path of the penetration of renewables from government policies and sector regulations and incentives in the fiscal, financial and regulatory realms.

Renewables generally include generation from wind, solar, biomass, geothermal, tidal and wave resources and the fraction of co-generation that uses gas. Mostly it excludes

Table 7.1 Share of electricity from clean energy sources in gross consumption (%) for the EU and selected countries

	2004	2005	2006	2007	2008	2009	2010	2011	2012	2013	2014
EU 28+	**14.4**	**14.9**	**15.4**	**16.1**	**17.0**	**19.0**	**19.7**	**21.7**	**23.5**	**25.4**	**27.5**
Denmark	23.8	24.6	24.0	25.0	25.9	28.3	32.7	35.9	38.7	43.1	48.5
Germany	9.4	10.5	11.8	13.6	15.1	17.4	18.1	20.9	23.6	25.3	28.2
Spain	19.0	19.1	20.0	21.7	23.7	27.8	29.8	31.6	33.5	36.7	37.8
France	13.8	13.7	14.1	14.3	14.4	15.1	14.8	16.3	16.4	16.8	18.3
Italy	16.1	16.3	15.9	16.0	16.6	18.8	20.1	23.5	27.4	31.3	33.4
Netherlands	4.4	6.3	6.5	6.0	7.5	9.1	9.6	9.8	10.4	10.0	10.0
Austria	61.8	62.4	62.4	64.6	65.2	67.8	65.7	66.0	66.5	68.0	70.0
Portugal	27.5	27.7	29.3	32.3	34.1	37.6	40.7	45.9	47.6	49.1	52.1
Finland	26.7	26.9	26.4	25.5	27.3	27.3	27.7	29.4	29.5	30.9	31.4
Sweden	51.2	50.9	51.8	53.2	53.6	58.3	56.0	59.9	60.0	61.8	63.3
Norway	97.3	96.8	100.2	98.5	99.6	104.7	97.9	105.5	104.4	106.9	109.6
UK	3.5	4.1	4.5	4.8	5.5	6.7	7.4	8.8	10.7	13.8	17.8

Source: Eurostat (2016) and International Energy Agency (IEA, 2016).

a large contribution from hydroelectric power. The term 'clean energy' often includes hydroelectric power, according to IEA (2016). In the present investigation, hydroelectric, wind and solar PV power are considered, with an emphasis on wind and solar PV power. The study includes EU economies with data available (part of the EU 28, plus Norway).

Countries in the EU started to use renewables early, in the 1990s, with other OECD countries not in the EU, notably Australia, New Zealand, Japan and the USA following. Large developing countries such as China, to a much lesser extent India, and other countries in Latin America became active in the use renewable sources of energy much more recently. Table 7.1 and Figures 7.1–7.5 provide a visualisation of these dynamics.

According to IEA (2016), several countries have had a high level of clean sources of energy since the beginning of the period of investigation (2004). Overall, 14.4% of the electricity produced in the EU and Norway was clean in 2004, climbing to 27.5% by 2014, but clean energy penetration is heterogeneous across the region. Since the beginning of the study period, Austria, Norway and Sweden have had clean energy shares above 50%. Denmark, Germany, Spain, Italy, Portugal and Finland were also above the European average in 2014. All countries doubled their penetration levels in the study decade, except for Spain, France and Austria (Austria had one of the highest levels since 2004), and the Nordic countries. Some countries more than doubled their renewable penetration, such as Germany and the UK. Finally, some of the more pronounced growth began around 2009–2010. Separated by energy source, Figures 7.1–7.5 show the dynamics in the selection of countries and regions.

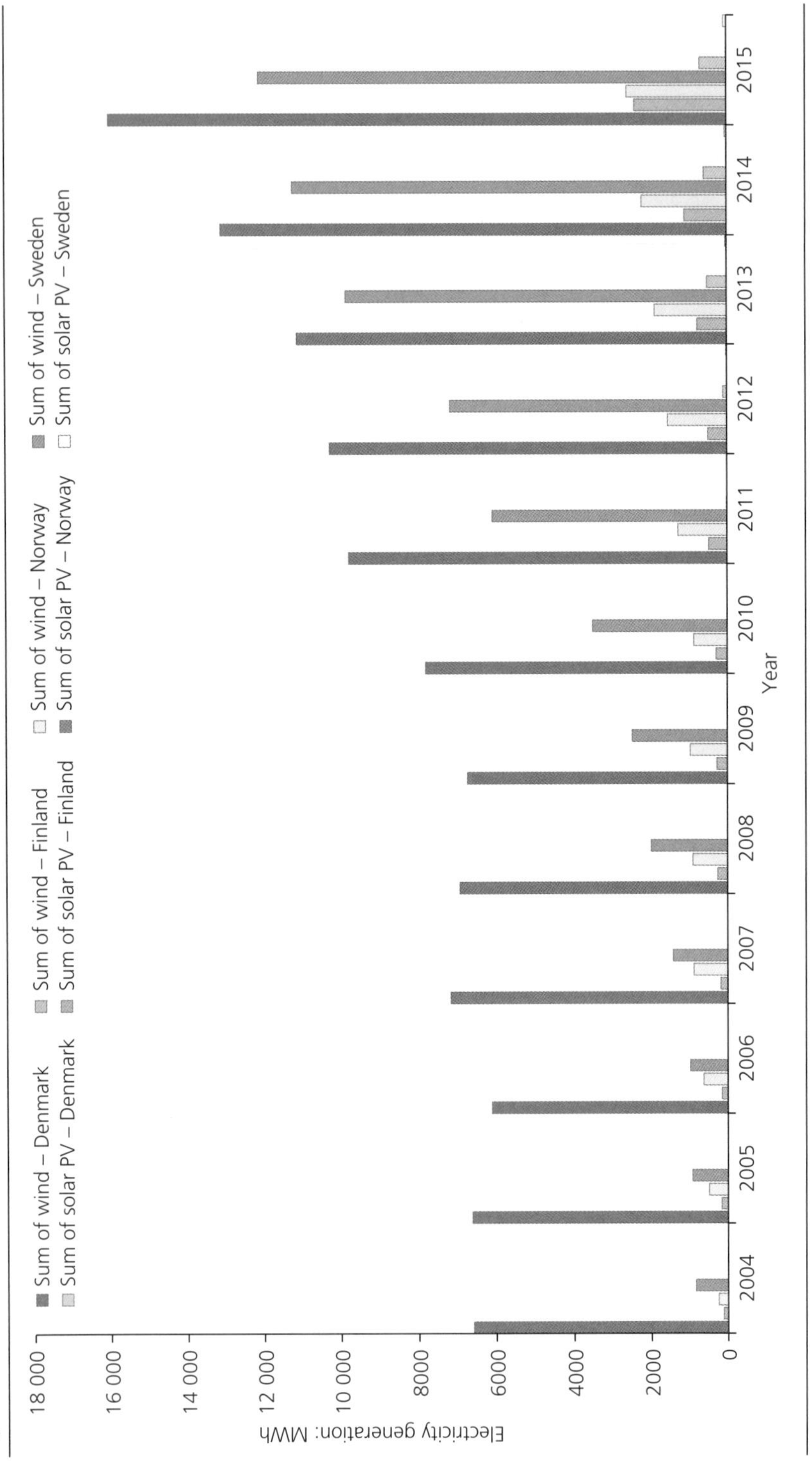

Figure 7.1 Wind and solar PV generation in the Scandinavian region. (Source: calculations on data from Eurostat, 2016) and IEA, 2016

Figure 7.2 Wind and solar PV generation in selected European countries. (Source: calculations on data from Eurostat, 2016)

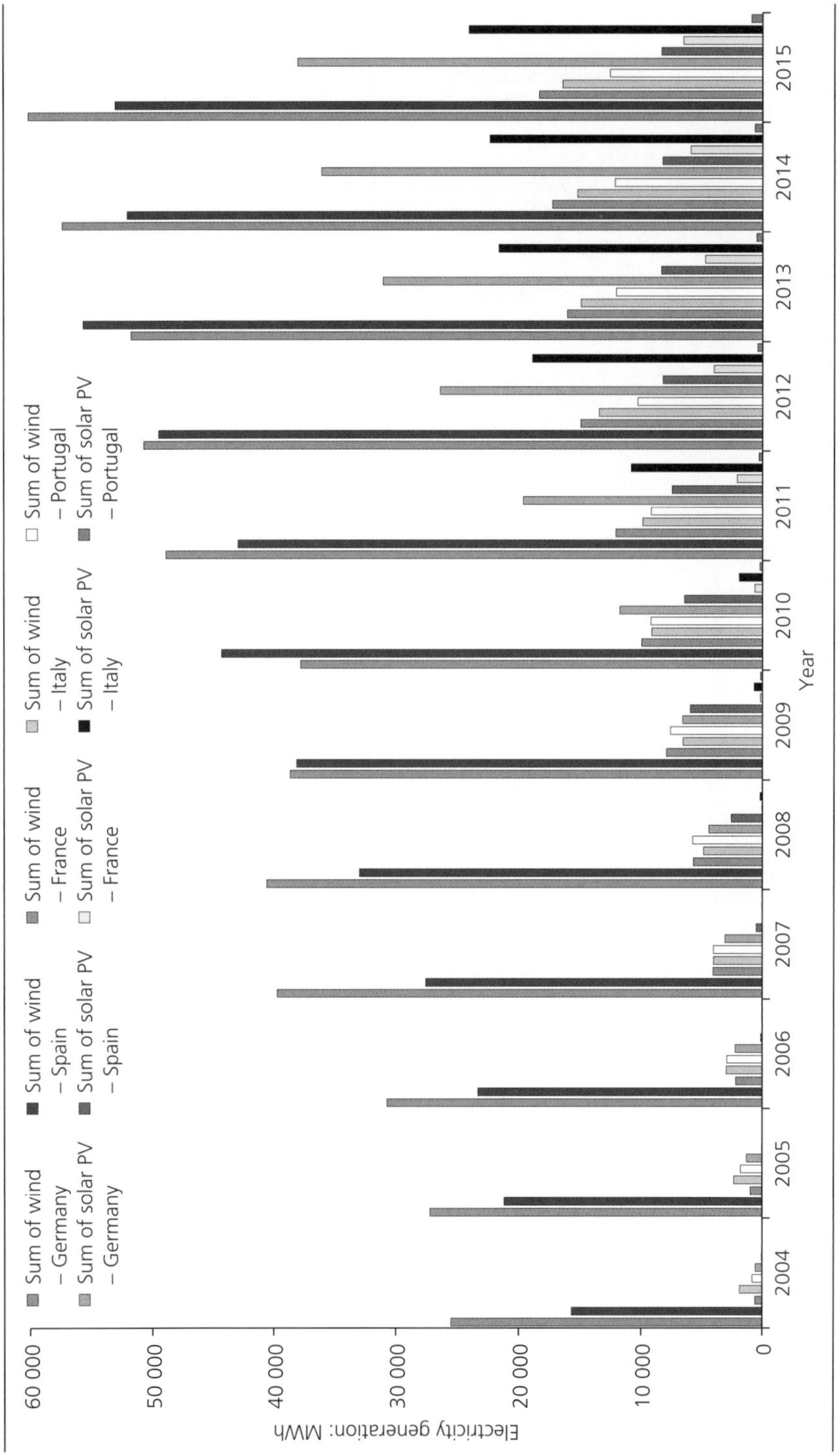

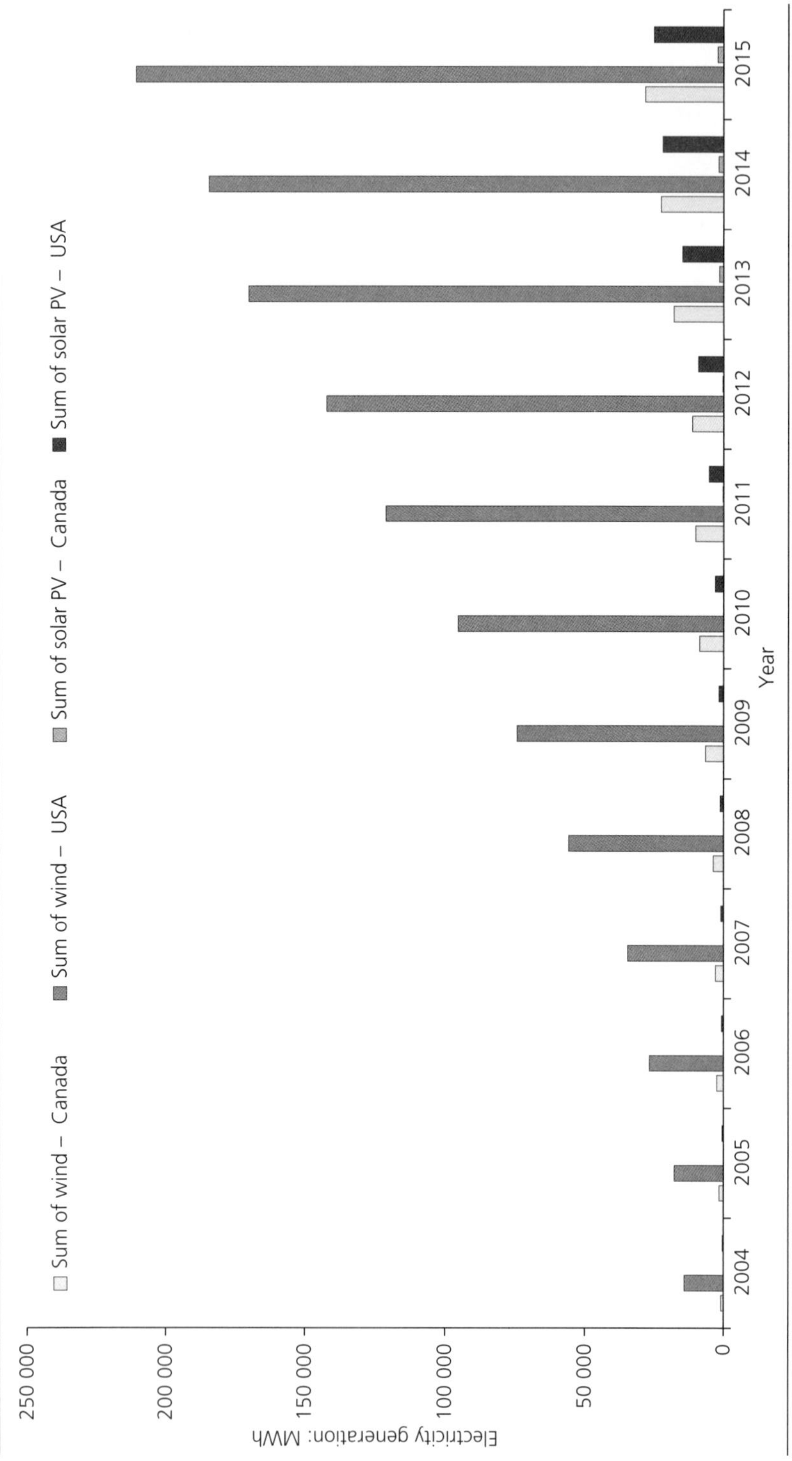

Figure 7.3 Wind and solar PV generation in North America. (Source: Own calculations with data on Eurostat, 2016)

Figure 7.4 Wind and solar PV generation in selected Asian Countries. (Source: calculations on data from Eurostat, 2016)

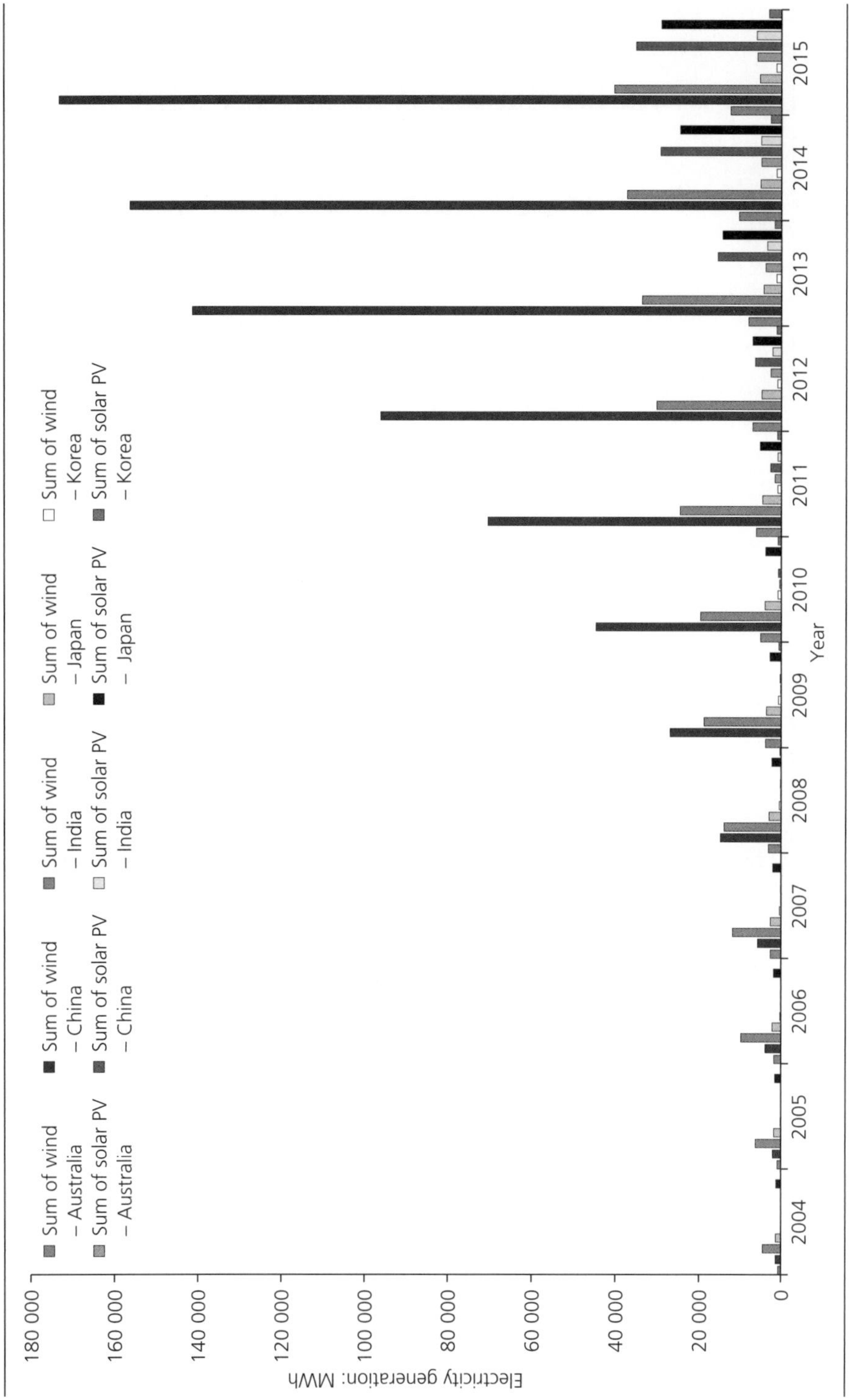

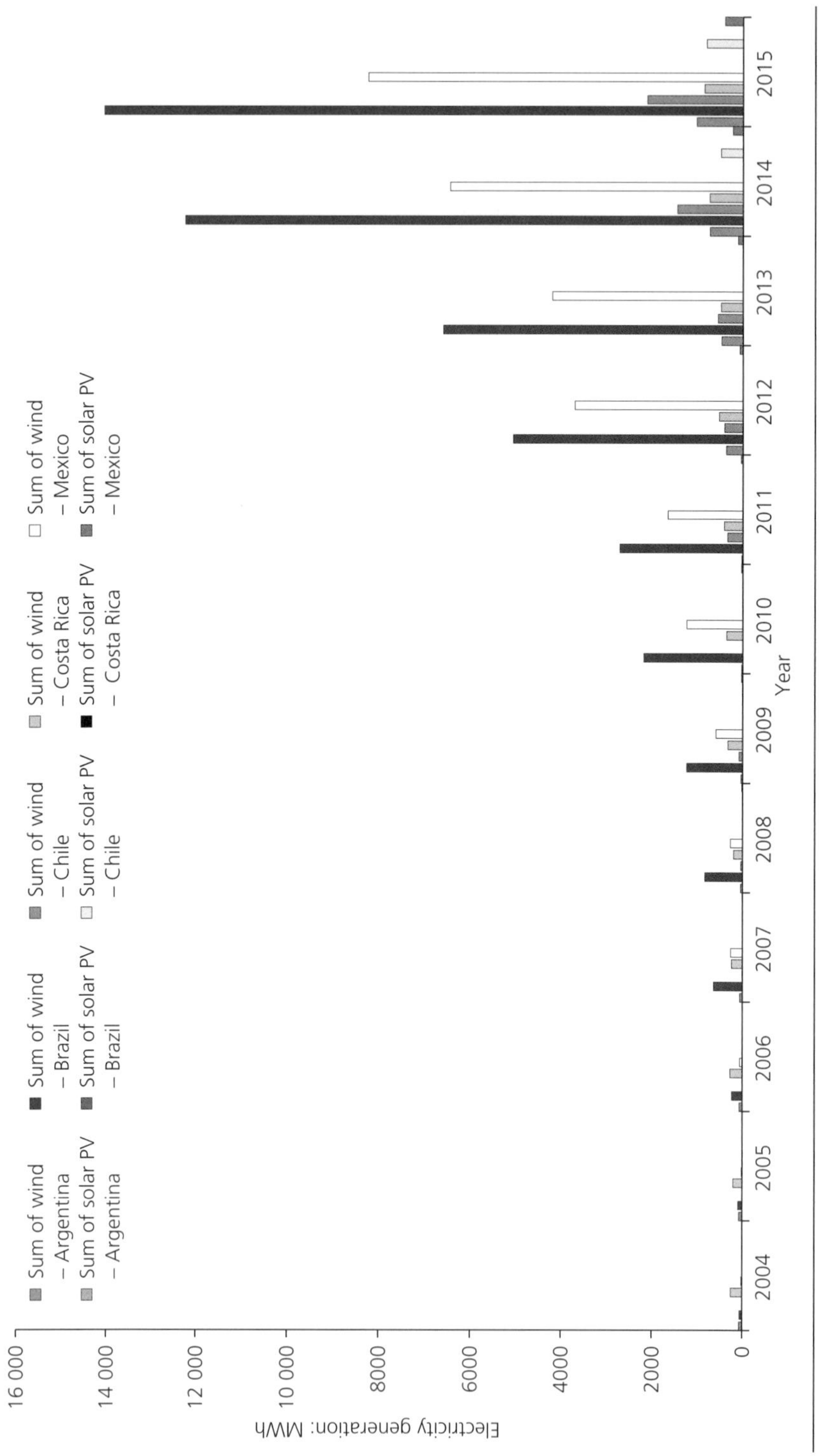

Figure 7.5 Wind and solar generation PV in selected Latin American countries. (Source: calculations on data from Eurostat, 2016)

The leading uptake of Denmark's wind generation is very noticeable, followed by Sweden (Figure 7.1). However, both Finland and Norway experienced more recent dynamic growth in wind energy, whereas Sweden moved to increase solar PV generation in the later years. The rest of the region had not embarked on solar PV energy production.

For Europe overall, selected countries also embarked on dynamic growth in wind energy production, with notable maturity in Germany and Spain (Figure 7.2). Note, however, the rather dynamic growth in solar PV generation in Germany and Italy, whereas others have stable growth rates. There are many implications, then, for new technologies for distributed energy and smart grids in specific urban projects. Not shown in Figure 7.2 is the rather dynamic growth in renewable sources in other economies in Central and Eastern Europe (IEA, 2016).

In North America (excluding Mexico), the USA is a leading producer of wind energy, generating several times the energy from wind of any European country, and Canada has increased its investment and production of wind energy since 2009 (Figure 7.3). Also notable is the solar PV uptake in the USA since about 2010. This is a time when new tax incentives were introduced – not for the entire country but for solar farms in localities where utility levels suggested investment in infrastructure to bring energy from distant areas of production to urban areas of high demand and load.

Notable in the Asian region is China's leadership in wind production, and the country's accelerated growth in this since the beginning of the study decade (Figure 7.4). Second in the region for the growth in wind production is India. In the case of Australia, solar PV generation has shown notable growth. China began solar PV generation around 2012. This dynamic but asymmetric growth will be useful for the analytical data analysis in the next section, and might have implications for policies towards distributed energy projects in urban areas and small communities.

In the case of selected countries in Latin America (Argentina, Brazil, Chile, Costa Rica and Mexico), Brazil, followed by Mexico, stand out as very dynamic in their wind energy production uptake, while Chile showed increased wind energy production in the later years (Figure 7.5). Again, the different rates of growth are correlated in the expanded sample with their government determinants, country size, and the different dynamics for renewable energy sources.

To close this section, a time series analysis was carried out for hydroelectric sources. Growth rates in this technology were found to be rather stable, with low growth in all the regions analysed. The selection of examples and data sets for the analysis of the countries discussed above depended on when the renewable energy uptake was clearly evident.

7.4. A model for government regulatory quality and penetration of renewables

A critical dimension for the current study is whether sectoral (energy) regulators are replicated across countries, or whether a sectoral regulator sets the endowments, administrative capabilities, instruments or policies of the regulated part of their sectors, similar to delegated regulators in other economic sectors. To account for this, the World Bank's Worldwide Governance Indicators can be used, which measure various aspects of governance (World Bank, 2016): (*a*) Voice and Accountability; (*b*) Political Stability and Absence of Violence; (*c*) Government Effectiveness; (*d*) Regulatory Quality; (*e*) Control of Corruption; and (*f*) Rule of Law. Two of the dimensions of governance were used in the analysis in this chapter – Government Effectiveness and Regulatory Quality. Each indicator was standardised to a normal distribution with a zero mean and a standard deviation of 1 (ranging from approximately −2.5 to + 2.5) for a large number of developed and developing countries (World Bank, 2016).

7.4.1 Panel data models applied to regulatory quality and production with renewables

Panel data analysis is a powerful methodological approach to both the time dimension and for comparison across countries. The following panel data model is very useful to account for both the time dimension of renewable uptake and the diversity of country cross-section cases for both developed and developing economies (and eventually city promotion policies)

$$\log(y) = f(x_1, x_2, x_3, z) \tag{1}$$

where the logarithm of the dependent variable (electricity generation in GWh) accounts for the launch and explosive behaviour in the investment in and production (not only capacity) by two renewable variables of interest: wind and solar PV power. (Note: capacity as an alternative dependent variable, related more to demand characteristics and reserves, could not be found for all countries analysed.) The x_1 vector includes variables that account for the regulatory impulse (divided into a government efficiency variable and, alternatively and more specifically, regulatory quality variable). Then, x_2 is a categorical variable of the market power of incumbents after regulatory changes have been enacted or there are similar but different competition environments; this ranges from high concentration (1) to market competition (7). x_3 represents structural changes in renewables, such as the break year in which each country began producing electricity with renewable energies in a clearly explosive way (represented as dummy variables taking the value 0 before the break year and 1 afterwards). It accounts for the endogeneity of renewable energy efforts. Finally, z denotes control variables (GDP per capita in purchasing power parity in US dollars ($ PPP) and a dummy variable for developed versus developing economies). The panel includes $j = 34$ countries (18 developed, and 16 developing) where data exist for regulatory compacts and where renewable uptake was clearly observable. The time frame for the panel data analysis extended from $t = 2004$ to 2015, the most recent year where comparable data exist. Thus, the panel consisted of 408 observations.

Panel data can be investigated using linear and non-linear time series and cross-section models and extensions, originally used for micro-econometric methods. Cameron and Trivedi (2005) discuss alternative approaches: (*a*) pooling data, (*b*) weighted least squares based on error variances, (*c*) demeaned panels, (*d*) panel in first differences and (*e*) generalised least squares with fixed effects.

The countries in the panel analysis were Argentina, Australia, Bolivia, Brazil, Canada, Chile, China, Colombia, Costa Rica, Denmark, El Salvador, Finland, France, Germany, Guatemala, Honduras, India, Italy, Japan, South Korea, Mexico, the Netherlands, Nicaragua, Norway, Panama, Paraguay, Peru, Portugal, Spain, Sweden, UK, USA, Uruguay and Venezuela. Ecuador was initially considered, but dropped from the analysis because it did not show any output in renewables. China was an outlier due to its strong renewables uptake with government indices that did not improve, meaning that incentives were not market driven. So, although China was also initially considered, it too was excluded from the analysis.

Table 7.2 provides descriptive statistics for the data.

From the basic statistics above, the mean, median and extreme values of the regulatory variables for Government Efficiency (ge) and Regulatory Quality (rq), measured in standard form by the World Bank statistics, were obtained, as well as limits to the variable of the market dominance proxy (mktdom), from a value of 1 for strong incumbent dominance to 7 for atomised and competitive markets, as measured by the World Economic Forum (Schwab, 2016). Note that the minimum value for hydropower

Table 7.2 Basic statistics for observations 1:01–12:34 (34 countries and 12 years) (missing values ignored)

Variable	Mean	Median	Minimum	Maximum
ge	0.641985	0.460904	−1.22895	2.35659
rq	0.642916	0.612308	−1.86000	1.92092
mktdom	4.39260	4.38000	2.51000	6.35000
hydro (GWh)	79933.2	25370.0	13.0000	1.10100e + 006
wind (GWh)	9739.07	926.500	0.00000	210100.0
solarpv (GWh)	1739.04	4.00000	0.00000	38000.0
breakyr	0.475490	0.00000	0.00000	1.00000
dummdevel	0.470588	0.00000	0.00000	1.00000
l_wind	7.37747	7.84065	0.00000	12.2553
l_solarpv	5.10154	4.77912	0.00000	10.5453
wind_1	9812.32	958.000	0.00000	210100.0
l_wind_1	7.39809	7.86749	0.00000	12.2553
l_solarpv_1	5.09965	4.76217	0.00000	10.5453
d_ge	−0.00437129	−0.00585917	−2.52445	2.90178
d_rq	−0.00819443	0.00257235	−2.06265	2.69694
d_mktdom	−0.00898990	−0.0400000	−2.52000	3.45000
d_ycapppp	−62.1069	632.220	−51979.2	37992.5
d_l_wind	0.163084	0.192128	−8.29761	5.32788
d_l_solarpv	0.243810	0.287682	−8.33086	4.11292
intge	0.397659	0.00000	−0.958573	2.25827
intrq	0.397629	0.00000	−1.07892	1.91157

Table 7.2 Continued

Variable	Standard deviation	Coefficient of variation	Asymmetry	Excess of kurtosis
ge	1.00403	1.56395	−0.0602117	−1.36447
rq	0.908813	1.41358	−0.372050	−0.825433
mktdom	0.975203	0.222010	0.0559991	−1.16350
hydro (GWh)	149518.0	1.87054	3.47888	15.0556
wind (GWh)	24608.8	2.52682	4.84263	27.8436
solarpv (GWh)	5460.65	3.14003	4.32932	19.6671
breakyr	0.500012	1.05157	0.0981572	−1.99037
dummdevel	0.499747	1.06196	0.117851	−1.98611
l_wind	2.65547	0.359943	−0.797893	0.441361
l_solarpv	2.94645	0.577562	0.0941527	−1.21408
wind_1	24871.9	2.53476	4.82004	27.3959
l_wind_1	2.63803	0.356583	−0.798912	0.447775
l_solarpv_1	2.94760	0.578000	0.107728	−1.20710
d_ge	0.429456	98.2448	0.459226	20.9459
d_rq	0.359556	43.8780	0.235745	21.8840
d_mktdom	0.679707	75.6078	0.849927	4.66310
d_ycapppp	6919.75	111.417	−3.92538	28.9796
d_l_wind	0.975959	5.98438	−3.15449	28.9217
d_l_solarpv	1.21412	4.97979	−3.71661	23.0250
intge	0.743551	1.86982	0.951108	−0.280037
intrq	0.687885	1.72996	0.923236	−0.308766

Source: generated from the data using the computational package Gretl.
l_, logarithms of a variable; _1, variables with one-period lag; d_, variables measured as first differences.
The modified variables were used in the alternative econometric specifications.

(hydro), solar PV power (solarpv) and wind power (wind) (all in GWh) is zero, meaning that for some countries and years, no generation from these sources occurred, before explosive growth began as a structural change in commitment to renewables that could then be inserted into smart city analyses (Sivaraman and Moore, 2012).

The same can be noted for the analysis of the standard deviation, asymmetry and kurtosis. Other variables in Table 7.2 include GDP per capital in $ PPP (ycapppp), and dummy variables for development levels by the two subgroups of countries (dummdevel). To account for the endogenous explosive growth of solar PV and wind generation, logarithms of the two variables were used (l_solarpv and l_wind, respectively). The results of the panel data analysis are shown in Tables 7.3 and 7.4.

7.4.2 Analysis of the results

Starting with a base model of pooled data (POLS), corrected for heteroskedasticity and serial correlation (*C*), the analysis yields weak significant effects for both ge and rq,

Table 7.3 Panel data analysis; 326 original observations; $j = 34$, $t = 12$

(a) Dependent variable: l_wind with ge

Variable	POLS (no heteroscedasticity)	WLS (based on error variance)	FE (time dimension)
C	−0.427	0.373	−1.069
	(−0.371)	(0.53)	(−1.39)
mktdom	1.2961***	0.997***	1.085***
	(3.65)	(6.33)	(5.85)
ycapppp	3.314e − 06	4.983e − 05**	1.306e − 04***
	(0.05)	(2.77)	(7.19)
breakyr	1.8341***	1.523***	1.447***
	(3.22)	(7.07)	(5.75)
dummdevel	2.5307*	2.202***	
	(1.92)	(4.76)	
ge	−0.841	−1.188***	−1.030**
	(−1.10)	(−4.84)	(3.43)
Adjusted R^2	0.422	0.520	0.464
log(likelihood)	−689.94	−457.03	−658.51

(b) Dependent variable: l_wind with rq

Variable	POLS (no heteroscedasticity)	WLS (based on error variance)	FE (time dimension)
C	−0.190	0.602	−1.019
	(−0.17)	(0.85)	(−1.02)
mktdom	1.244***	0.989***	1.047***
	(3.80)	(6.17)	(5.99)
ycapppp	1.262e − 05	3.658e − 05*	1.562e − 04***
	(1.21)	(1.97)	(9.31)
breakyr	1.785***	1.718***	1.396***
	(3.75)	(7.76)	(5.76)
dummdevel	2.646*	2.087***	
	(1.84)	(4.33)	
rq	−1.211	−1.025***	−1.700***
	(−1.92)	(−4.01)	(−5.87)
Adjusted R^2	0.437	0.464	0.459
log(likelihood)	−684.16	−459.22	−471.33

Significance: at 1%, ***; at 5%, **; at 10%, *. In all estimations standard errors are corrected for heteroskedasticity and serial correlation.

Table 7.4 Panel data analysis; 326 original observations; $j = 34$, $t = 12$

(a) Dependent variable: l_solarpv with ge

Variable	POLS (no heteroscedasticity)	WLS (based on error variance)	FE (time dimension)
C	−3.197	−2.629***	−4.180***
	(−1.74)	(−2.87)	(−4.05)
mktdom	1.103**	0.761***	1.034***
	(2.91)	(3.58)	(4.67)
ycapppp	1.389e − 04**	1.661e − 04**	1.8.8e − 04***
	(2.29)	(7.86)	(6.64)
breakyr	2.563***	2.233***	2.475***
	(5.53)	(7.36)	(5.75)
dummdevel	1.355	2.025***	
	(0.76)	(3.23)	
ge	−3.196**	−3.435***	−2.264**
	(−2.68)	(−9.10)	(−5.43)
Adjusted R^2	0.417	0.482	0.421
log(likelihood)	−519.81	−328.89	−65.21

(b) Dependent variable: l_solarpv with rq

Variable	POLS (no heteroscedasticity)	WLS (based on error variance)	FE (time dimension)
C	−3.136	−2.073***	−4.752***
	(−0.17)	(−2.65)	(−4.75)
mktdom	0.859*	0.647***	0.918***
	(3.80)	(3.68)	(4.38)
ycapppp	2.031e − 04	2.111e − 04***	2.505e − 04***
	(0.21)	(11.41)	(8.06)
breakyr	2.472***	1.952***	2.353***
	(3.73)	(7.24)	(7.76)
dummdevel	1.085	1.719***	
	(1.84)	(3.33)	
eq	−4.073*	−4.534***	−3.370***
	(−1.92)	(−11.99)	(−7.01)
Adjusted R^2	0.459	0.609	0.504
log(likelihood)	−512.19	−457.03	−658.51

Significance: at 1%, ***; at 5%, **; at 10%, *. In all estimations standard errors are corrected for heteroskedasticity and serial correlation.

applied to wind power (non-significant ge; only rq is significant at the 10% level), with a stronger effect of these conditioning variables on the production of solar PV energy, where the more direct rq effect more strongly promotes or hinders solar PV projects in the sample of countries. An explanation for this is that regulatory upgrades could trigger stronger production dynamics in solar PV generation. However, a preliminary analysis of the dynamics of both ge and rq for all countries in the sample (not presented here) indicated that developed countries generally show positive government efficiency and regulatory quality indexes, but have seen them decline between 2004 and 2015. For developing economies, the indexes are generally negative and even declining, so the trend of the indexes might be more important than the sign, and hence some significant results might show them with negative signs.

In the weighted least squares approach (WLS) based on error variances, to counteract for possible statistical inefficiency and inconsistency, the models are much clearer in all regressors for both logarithm variables (l_wind, and l_solarpv). For wind power, ge, a general characteristic, is much more significant than the more direct rq effect, while for solar PV power, the more direct rq shows stronger effects. This applies to both developed and developing economies, as shown by the highly significant dummy variable (dummdevel) in all cases. The lowering of the market dominance variable (mktdom) is also very significant and with the correct sign in all cases, meaning that a regulation that opens the market for competition generates incentives for increased wind and solar output, according to the theoretical argument. This is irrespective of the perceived ge or rq effect. In other words, a regulation could be less than complete, but if it moves towards market-driven activities, competitors and investors would react positively. Extension towards the analysis and planning of smart cities would imply that many actors are needed as investors in the uptake of renewables projects, in a market-prone economic environment. To finish the econometric analysis, control variables for economy size and structural change (GDP per capita in $ PPP, and the break year of expansion of renewables output) are significant in all cases for the WLS panel.

Further, for the case of a model with fixed effects (FE) in the time dimension, panel data models for both wind and solar PV generation are very significant, including the entire sample (no separation of developed from developing economies). Again the variable rq is overall more significant and stronger than ge, while the rest of the variables of the FE models with generalised least squares in the panel are significant at the 1% level (Cameron and Trivedi, 2005). rq was statistically more prevalent than the overall ge index. Even in cases where rq was negative, it showed improvements for most developing countries.

7.5. Dynamics of renewable output, "ge" and "rq": a robustness check for interaction terms

Let us observe how regulatory packages for renewable electricity production are created and implemented, since there is much variation in regulatory instruments and long-term outlook in energy policy and the coexistence of energy policy and climate change. What is the interrelationship between the break year of explosive output in the respective economies, and the regulation perspective (for both ge and rq)? A series of statistical runs with the interaction terms (intge, intrq) would explore such a potential endogeneity, providing a robustness test for the earlier analyses. No significant changes were found from a series of additional runs, meaning that perceptions of ge and/or rq persisted as explanatory variables for wind and solar penetration in the countries considered, with different effects for the dummdevel variable. The above could then be applied to delegated government effectiveness and efficiency at a local or city level.

A test for the output level of renewables and the level of development to determine the speed of increase in wind and solar PV production would account for how development levels affect the output speed of l_wind and l_solarpv via an interaction term (devspeed), to account for development and renewable dynamics rather than development levels and the regulation variables ge and rq as above.

First, for wind output with respect to the regulatory variables ge, and rq, the test for two panel data models shows a clear significant and relatively stronger impact of the closer rq than the overall ge. The same occurs for the case of the solar PV set of equations. Second, the interaction of the development-level dummy and the speed of output is highly significant in all models. When an additional lagged dependent variable is added to the dynamic models, the lag is highly significant but with a larger effect in the case of wind than solarpv. The interaction term devspeed (the development-level dummy multiplied by the uptake of wind and solar PV power) shows high significance in all tested models and accounts for the state of development (Sivaraman and Moore, 2012).

In a final exercise, the 18 developed economies in the panel study were removed, and another panel analysis carried out only on the 16 developing countries in the subsample, even though the degrees of freedom would make conclusions rather tentative. The results demonstrated that emerging markets show statistical significance of the overall ge again for wind power, but given that solar PV uptake is rather marginal in the sample period, no significance was present in the emerging economies subsample. The results applied to OECD developed countries were maintained and similar to the overall sample of 34 countries.

The implications of the analysis are as follows.

- Wind and solar PV generation are new technologies that have become key in the shift of the energy matrix for the 34 countries analysed, both developed and developing, where cost reductions have impelled fast growth, and dynamic breaking years for the deployment of solar PV resources. Moreover, since its launch in about 2005 of solar PV energy has seen exponential growth. Distributed energy projects using solar and wind sources of energy are often a characteristic of overall smart city planning so that the explosive growth rate of renewable energy observed in the set of countries analysed clearly relates to smart city governance and commitment.
- The dynamic character of renewables was integrated into the modelling exercise using panel analysis via the logarithms of the dependent variables, while hydropower did not react to all of the explanatory variables, such as the level of

development, and the perception indices ge and rq. This finding signals a need to deepen market-oriented regulatory changes for those parts prone to deregulation. Such an approach coincides with smart city aims where communities and agents actively participate in the design of specific innovations. Such issues are considered further in Chapter 11.

- Accounting for the dynamics as a robustness check was tried for the original models by analysing the interaction terms between ge/rq and the impulse of the trigger break year of exponential growth in wind and solar PV production (in GWh). However, the impulse of the break year is so strong that the interaction term was not very significant. This is particularly the case for China.
- The market dominance of the incumbents (mktdom) was significant throughout the analysis; however, its negative impact is reduced as renewables become increasingly present, and where market-driven energy models change the face of communities.
- The level of economic development also helps explain the time when wind and solar PV production began, with most Latin American economies in the sample, plus China and India, beginning much later than their developed counterparts. Since other developing economies faced little or no uptake in renewables in the sample analysed, no additional analysis was done separating the two subsamples of developed and developing countries as a further robustness check, for insufficient degrees of freedom.
- In the analysis it seems apparent that the market perception of rq has more explanatory power than the overall ge, mostly in solar PV investment and output. It is possible that this finding has to do with the fact that solar PV projects are more recent than well-established wind projects. Some cases where the sign of the indexes is negative could explain why ge/rq were declining for a subset of developed economies, while some developing countries showed negative indexes, although improving. The above implies that the direction of improving perceptions of ge/rq are more important than simply their levels in the data.
- Finally, the modelling exercise of the extended dynamic models shows that some versions of panel data analysis, such as the WLS and FE panel analyses, render better goodness-of-fit tests, including lagged dependent variables, meaning that development speeds should be subject to strong but clear policies friendly to new investment and output.

7.6. Conclusions and implications for policy

The research considered in this chapter investigated a panel of 34 countries and a range of 12 years to test the importance of government efficiency and regulatory quality indices in the promotion of investment and the output of renewable wind and solar PV energy, along with the variable of market dominance persistence that significantly affects the decision to invest and produce. A first comment on the analysis is that moving from the country level to other instances at more granular, local levels presupposes that the process of institution building and delegation for governance indices can be replicated and that similar experiences apply to both the country level and localities. Since government effectiveness and regulatory quality do not yet exist at local levels, the country-level analysis could serve as an overall framework for the treatment of policy for local incentives for renewables.

The findings also complement theoretical arguments that decisions to change the energy mix across the world to increase renewables should result in policies that incentivise markets and the take-off of wind and solar PV projects, where perceptions of government effectiveness and regulatory quality are clear for investors, meaning ensuring better governance along with less market concentration. Moreover, the existence of market-driven renewables implies active market and technology agents are needed for visionary smart city plans and projects. Now, from the data panel analysis, in its alternative forms, simpler models have relatively lower explanatory power than more involved GLS and GLS with FE models that account for greater dynamics. After inclusion of the dynamic behaviour of both the dependent and explanatory variables in the panel data analysis, via interaction terms for robustness, the latter models seem to improve the explanation and fitness tests that directly test the endogeneity.

Both the ge and rq indices, plus an index of market dominance after regulatory changes, help explain the explosive growth of wind and solar PV output across a group of 18 developing and 16 developed countries – a measure that is new to the literature. However, as an extension of the present study, other analyses could further study specific regulatory instruments (regulatory, fiscal and financial) to incentivise investment projects in renewable and clean energy sources, to change the energy mix and to apply them to urban communities. In the main, these sources of incentivising investment are applied at the aggregate, country level in the present research (regulatory, fiscal and financial, as mentioned). Finally, case studies comparing developed mature markets for renewable wind and solar PV investment and production could then be contrasted to the recent dynamics of renewable uptake investment and production in the main deregulated and market-oriented developing countries. In closing, energy dynamics seem to be a relevant complement to enhance smart city projects with agents in all urban ecosystems.

Acknowledgements

The author wishes to acknowledge and thank two reviewers for this research. All errors are the author's.

REFERENCES

Adelman DE and Spence DB (2015) *Cost–Benefit Politics in U.S. Energy Policy*. Kay Bailey Hutchinson Center for Energy, Law, and Business, University of Texas at Austin, TX, USA. https://repositories.lib.utexas.edu/handle/2152/33365 (accessed 20/06/2019).

Aguirre M and Ibikunle G (2014) Determinants of renewable energy growth: a global sample analysis. *Energy Policy* **69**: 374–384.

Albino V, Berardi U and Dangelico DR (2015) Smart cities: definitions, dimensions, performance, and initiatives. *Journal of Urban Technology* **22(1)**: 3–21.

Alchian D and Demsetz H (1972) Production, information costs, and economic organization. *American Economic Review* **62**: 777–795.

Ata NK (2016) An evaluation of renewable energy policies across EU countries and US states: an econometric approach. *Energy for Sustainable Development* **31**: 83–90.

Basher SA, Masini A and Aflaki S (2015) Time series properties of the renewable energy diffusion process: implications for energy policy design and assessment. *Renewable and Sustainable Energy Reviews* **52**: 1680–1692.

Borenstein S, Bushnell JB and Wolak F (2002) Measuring market inefficiencies in California's restructured wholesale electricity market. *The American Economic Review* **92(5)**: 1376–1405.

Bunn DW and Muñoz JI (2015) Supporting the Externality of Intermittency in Policies for Renewable Energy. *Energy Policy* **88**: 594–602.

Bushnell J and Stoft SE (1997) Improving private incentives for electric grid investment. *Resource and Energy Economics* **19**: 85–108.

Cameron AC and Trivedi PK (2005) *Microeconometrics: Methods and Applications.* Cambridge University Press, New York, NY, USA.

Chourabi H, Nam T, Walker S *et al.* (2012) Understanding smart cities: an integrative framework. *45th Hawaii International Conference on System Sciences, Maui, HI, USA.* https://ieeexplore.ieee.org/document/6149291 (accessed 20/06/2019).

Coen D and Thatcher D (2008) Network governance and multi-level delegation: European networks of regulatory agencies. *Journal of Public Policy* **28(1)**: 49–71.

EERE (Energy Efficiency and Renewable Energy) (2012) *Tax Credits for Energy Efficiency.* US Department of Energy, Washington, DC, USA. https://www.nrel.gov/docs/fy14osti/60197.pdf (accessed 20/06/2019).

Egerer J, Rosellón J and Schill WP (2015) Power system transformation toward renewables: an evaluation of regulatory approaches for network expansion. *Energy Journal* **36(4)**: 105–128.

Estache A and Martimort D (1998) *Transaction Costs, Politics, Regulatory Institutions, and Regulatory Outcomes.* The World Bank, Washington, DC, USA, EDI Regulatory Reform Discussion Paper.

Eurostat (2016) http://ec.europa.eu/eurostat/web/energy/data/energy-balances (accessed 20/06/2019).

Executive Office of the President of the United States (2016) *Incorporating Renewables into the Electricity Grid: Expanding Opportunities for Smart Markets and Energy Storage.* US Government, Washington, DC, USA. https://obamawhitehouse.archives.gov/sites/default/files/page/files/20160616_cea_renewables_electricgrid.pdf (accessed 20/06/2019).

Gabriel SA, Conejo AJ, Fukker JD, Hobbs BF and Ruiz C (2013) *Complementary Modeling in Energy Markets.* Springer, Berlin, Germany.

Griffin JM and Puller SL (2005) *Electricity Deregulation: Choices and Challenges.* University of Chicago Press, Chicago, IL, USA.

Harris C (2006) *Electricity Markets, Pricing, Structures, and Economics.* John Wiley, Chichester, UK.

Hogan W (2002) Electricity market restructuring: reform of reforms. *Journal of Regulatory Economics* **21**: 103–132.

Hogan W, Rosellon J and Vogelsang I (2010) Toward a combined merchant-regulatory mechanism for electricity transmission expansion. *Journal of Regulatory Economics* **38**: 113–143.

IEA (International Energy Agency) (2016) IEA Renewables Information Statistics. Archive 2016. http://www.oecd-ilibrary.org/energy/data/iea-renewables-information-statistics_renewab-data-en (accessed 20/06/2019).

Jacobsson S and Lauber V (2006) The politics and policy for energy system transformation – Explaining the German difussion of renewable energy technology. *Energy Policy* **34(3)**: 256–276.

Jordana J and Levi-Faur D (2005) The diffusion of regulatory capitalism in Latin America: sectoral and national channels in the making. *Annals of the American Academy of Political and Social Science* **598**: 102–124.

Kessides IN and Wade DC (2011) Towards a sustainable global energy supply infrastructure: net energy balance and density considerations. *Energy Policy* **39**: 5322–5334.

Laffont JJ and Tirole J (1991) Politics of government decision making: the theory of regulatory capture. *Quarterly Journal of Economics* **106(4)**: 1089–1127.

Li X, Paster M and Stubbins J (2015) The dynamics of electricity grid operation with increasing renewables and the path toward maximum renewable deployment. *Renewable and Sustainable Energy Reviews* **47**: 1007–1015.

Majone G (1997) From the positive to the regulatory state: causes and consequences of changes in the mode of governance. *Journal of Public Policy* **17(2)**: 139–167.

Majone G (2005) *Dilemmas of European Integration: The Ambiguities and Pitfalls of Integration by Stealth.* Oxford University Press, Oxford, UK.

Pollack M (2003) *The Engines of European Integration.* Oxford University Press, Oxford, UK.

Pollitt MG (2008) The arguments for and against ownership unbundling of energy transmission networks. *Energy Policy* **36(2)**: 704–771.

Pollitt MG (2012) Lessons from the history of independent system operators in the energy sector. *Energy Policy* **47**: 32–48.

Pollitt MG, Davies S, Price CW, Haucap J and Mulder M (2007) Vertical unbundling in the EU electricity sector. *Intereconomics Forum* **42(6)**: 292–310.

Rouhani OM, Niemeier D, Gao HO and Bel G (2016) Cost–benefit analysis of various California renewable portfolio standard targets: is a 33% RPS optimal? *Renewable and Sustainable Energy Reviews* **62**: 1122–1132.

Schwab K (ed.) (2016) *The Global Competitiveness Report 2015–2016*. World Economic Forum, Geneva, Switzerland. http://www3.weforum.org/docs/gcr/2015–2016/Global_Competitiveness_Report_2015–2016.pdf (accessed 20/06/2019).

Sivaraman D and Moore MR (2012) Economic performance of grid-connected photovoltaics in California and Texas (United States): the influence of renewable energy and climate policies. *Energy Policy* **49**: 274–287.

Spiller P (1990) Politicians, interest groups, and regulators: a multiple-principals agency theory of regulation, or let them be bribed. *Journal of Law and Economics* **33**: 65–101.

Tomain JP (2016) A perspective on clean power and the future of US energy politics and policy. *Utilities Policy* **39**: 5–12.

Vidal-Amaro JJ, Ostergaard PA and Sheinbaum-Pardo C (2015) Optimal energy mix transitioning from fossil fuels to renewable energy sources – the case of the Mexican electricity system. *Applied Energy* **150**: 80–96.

Wolak FA (2015) Measuring the competitiveness benefits of a transmission investment policy: the case of the Alberta electricity market. *Energy Policy* **85**: 426-444.

World Bank (2016) Worldwide Governance Indicators. http://info.worldbank.org/governance/wgi/index.aspx#home (accessed 20/06/2019).

ice Publishing

Nuttall, William J, Gibson, David V, Trzmielak, Dariusz and Ibarra-Yunez, Alejandro
ISBN 978-0-7277-6425-6
https://doi.org/10.1680/emsc.64256.139

Chapter 8
Future of community energy systems: lessons from agent-based modelling

J Richard Snape and Peter J Boait
Institute of Energy and Sustainable Development, De Montfort University, UK

8.1. Introduction

Smart energy usage forms a key component of any definition of a smart city. As highlighted in a recent Intergovernmental Panel for Climate Change report (IPCC, 2018), the urgency of energy transitions to low and zero carbon system grows ever greater, while United Nations Sustainable Development Goals (SDGs) (UN, 2018) emphasise the need for access to affordable and clean energy alongside sustainable cities and communities. Policy-makers, practitioners and academics recognise the need to accelerate transitions to clean energy systems, within which community energy schemes have a potentially large role to play. Community energy schemes can be understood in this city context as a combination of technology, infrastructure and business models (such issues have been considered in Chapters 2, 4 and 6), which are used to provide energy to a group of customers, who may also be investors, defined by shared locality. The scale of the locality could range from a few streets to the whole city. The value that such schemes can bring to a vision for the smart city stems from the increased trust and engagement between customer and supplier that can be engendered by a not-for-profit locally based organisation. Such issues will be considered further in Chapter 11.

However, community energy schemes as part of a smart city future face their own challenges. Understanding how models might be scaled up may be difficult, whether scale-up is by mass replication of very small schemes, likely to be highly diverse and administratively burdensome, or via a smaller number of larger schemes, which may lose the essence of community that works at smaller scale. In addition, schemes must exist in the context of differing and evolving technology ownership, for example with increasing electric vehicle (EV) ownership and changing modes of heating and cooling via use of heat pumps and building fabric changes.

This chapter proposes that agent-based modelling (ABM) is a highly useful technique to model the emergence and behaviour of community energy schemes in diverse contexts, and uses a number of implemented models to illustrate how the technique may be used to provide insight into smart community energy schemes that would not be possible using other modelling techniques. Use of ABM to understand the interactions between technology, economics, community members and policy is discussed, and the potential for analysing future community energy systems, including those within smart cities, is discussed.

The context for this chapter is the need for sustainable development of energy systems in a world with increasing population and urbanisation. The UN explicitly recognises affordable and clean energy as well as sustainable cities and communities as essential SDGs (UN, 2018; goals SDG 7 and 11). The IPCC (2018) has alerted us to the need for urgent and high impact action to decarbonise energy provision if we are to limit temperature change due to climate change, saying

> Pathways limiting global warming to 1.5°C with no or limited overshoot would require rapid and far-reaching transitions in energy, land, urban and infrastructure (including transport and buildings), and industrial systems (high confidence). These systems transitions are unprecedented in terms of scale.

Such challenging imperatives are a call to arms for those who wish to use developments in renewable energy provision alongside real-time data-processing capabilities to provide better and more sustainable living conditions. A recent review (Nerini *et al.*, 2018) of the synergies and trade-offs between SDG 7 (affordable access to clean energy) and the other 2030 goals note that

> For SDG7, there is a tension between the need for action required to rapidly address urgent energy-related issues (for example, energy access), and the careful planning of complex energy systems that underpin long-term development outcomes. Balancing these needs will require new skills and capacity to build country-specific and regional strategies.

These imperatives for energy transition arise at a time when populations continue to increase and urbanisation continues, particularly in developing countries where the migration to cities results in rapid increase in demand on city infrastructure through both formal and informal settlements.

How does this fit with the smart city? There is no doubt that the smart city is an active and attractive proposal, recognising the potential for integrated service delivery informed by (near) real-time information to provide sustainable services for city residents. An essential component of any smart city is smart provision of sustainably generated energy. One of the major issues with sustainable energy, particularly electricity generated from renewable sources, is that the time when electricity is produced is dictated by the weather, and this does not always correspond with the time when consumers wish to use that electricity. Thus, balancing supply and demand becomes more challenging than with conventional generation and much research focuses on ways to use smart systems combining real-time data, technology and consumer behaviour to assist this endeavour.

Intervening in such a complex system as energy supply is a daunting business. Even within a single city the system can service the interaction of millions of energy-using devices, thousands of consumers, hundreds of suppliers and dozens of regulatory bodies. This creates highly interdependent relationships, where an intervention may have multiple unintended consequences. This possibility is exacerbated in the presence of a drive for rapid transition. To attempt to mitigate the risk of such a transition, any greater *a priori* understanding of the potential consequences of policies or technological interventions is useful.

To achieve the desired rapid transition to clean sources of energy, many look to introducing community energy schemes to locally manage energy supply and demand, and this is the scheme that the remainder of this chapter explores. Community energy schemes can be an effective method to engage in rapid transition, with existing community relationships providing a framework that may be built upon to introduce measures that may otherwise be infeasible. However, research shows that community energy scheme participants are far from homogeneous and that trust within communities cannot be taken for granted (Walker *et al.*, 2010). With this in mind, a modelling technique that can incorporate differing community structures to assess potential for community energy schemes is highly desirable.

This chapter examines insight gained from a number of studies using ABM that illustrate the potential of the technique to provide useful models in challengingly diverse contexts. The potential of the technique has been recognised for some years, and it has been cited as an area for development in modelling energy systems to meet 21st-century challenges (Pfenninger *et al.*, 2014), particularly due to the technique being able to cope with complexity more readily than alternatives (Rai and Henry, 2016).

The chapter is organised as follows. Section 8.2 provides a short overview of community energy, and Section 8.3 summarises the ABM method and outlines its suitability for modelling community energy systems. Section 8.4 presents four case studies where ABM has been used to investigate community energy schemes, chosen to illustrate the different facets of community energy and contexts that may be investigated using this technique: Section 8.4.1 deals with renewable energy adoption in communities; Section 8.4.2 shows how ABM can be used to model smart community demand side management; Section 8.4.3 illustrates the use of ABM to model an integrated community scheme; and Section 8.4.4 uses ABM to analyse the effectiveness of a financial incentive policy. Section 8.5 draws together the case studies in an analysis of the uses to which ABM might be put in a community energy context, before Section 8.6 concludes.

Although the chapter ends with a (perhaps predictable) call for more work to be completed in the area, results from models already implemented are examined and their implications for future community energy systems discussed.

8.2. Community energy overview

The focus on community energy in this chapter arises because of a fundamental difficulty with conventional retail energy supply in the context of a smart city. This is the lack of trust between suppliers and consumers (Mumford and Gray, 2010). This impedes consumer confidence in the many changes necessary to the everyday procurement and use of energy that are inevitable in the transition to a smart city. The consumer resistance to the roll out of smart meters in the UK by the major energy retailers illustrates this point. The not-for-profit or cooperative governance expected for community energy, combined with localised brand names and leadership by individuals who are locally known, can improve the trust in this relationship.

However, the meaning of the term 'community energy scheme' is not settled. Many schemes have been cited as examples of community energy systems. In the early 2000s, 'community energy' often referred to community ownership of relatively large-scale renewable electricity generators, such as wind turbines. Examples of such schemes have been studied to understand the part that such schemes might play in future energy systems and examine whether evidence existed to confirm proposed benefits of community schemes (e.g. Toke, 2005; Walker *et al.*, 2010). Community ownership schemes remain a fundamental component of community energy, with community-owned solar and wind power installations ongoing in the UK, mainland Europe and globally. The potential benefit of such schemes in a city as elsewhere is that widespread local ownership can overcome the 'not in my back yard' resistance to renewable energy installations, and they can mobilise 'patient capital' for generator investments that might not otherwise be funded.

More recently, particularly with the advent of smart meters and smart grids, schemes based around supply contracts rather than asset ownership have been described as community energy schemes, especially in electricity supply and demand. Among these, many use distributed renewable generation resources (such as rooftop photovoltaic (PV) systems) to provide some or all of the energy required by consumers within a defined community. For example, in Oxford, in the UK, the SWELL trial of a particular community energy business model (Energy Local, 2016) used pre-existing awareness of investment-driven community energy from the county-wide Westmill cooperative to launch a more localised scheme to use real-time information about consumption within the community to use local renewable generation within the community as much as possible. This trial illustrates how the intensive support to consumers that can be provided by this kind of scheme can facilitate adoption and effective use of demand response technology and smart metering of a time-of-day (TOD) tariff (Boait *et al.*, 2017). The use of ABM to model this scheme and extrapolate from it is described in Section 8.4.3.

Community energy schemes are often seen by national policy-makers as an integral part of future energy systems. In the UK, policy refers to community energy schemes; however, support for schemes can appear sporadic, for example the Urban Community Energy Fund designed to support early growth of schemes in urban environments closed in 2016.

Community energy schemes in the UK have often been focused on electricity supply, and the work described in this chapter concentrates largely on the supply of electricity, rather than other forms of energy. However, in other locations – and increasingly in the UK – another form of community energy scheme that is in use and has potential for integration into a smart city future is community heating. Heating or cooling of buildings to provide thermal comfort for occupants often forms a very large component of building energy demand, as maintaining a temperature other than ambient is expensive energetically. Schemes for community district heating using a dedicated heat network exist, although their prevalence appears to be dependent on cultural mores, as they are common in northern Europe and certain areas of North America, but less so in the UK. District heating can be particularly suitable for the urban environment because geographical density of dwellings is helpful to avoid the large losses incurred in transporting heat. In warmer climates, district cooling is also possible, but appears to be relatively less common (Dominković *et al.*, 2017).

Increasingly, policy-makers are recognising the need to decarbonise heat. Thus, community district heating could form a part of many future smart cities. However, it seems that barriers exist in terms of market systems and cultural expectations in some countries. Coupled with the expense of installing large-scale infrastructure for the distribution of heat, it is perhaps more likely that in the UK, where typically heating is dealt with on a per-building basis, the current natural gas boilers will often be replaced individually by more sustainable heating sources, for example heat pumps. These too can be managed within a community scheme, as they present a significant electrical load that in effect couples the electricity system to a large heat store in the form of many dwellings with thermal mass. If managed in a smart way, this heat store may be used to provide flexibility via demand-side response to the electricity system. The integration of heat supply with electricity supply is a fertile area for further research, with early work suggesting that such integration will be essential (Boait *et al.*, 2017; Good and Mancarella, 2018).

8.3. Agent-based modelling

In the energy sector, ABM has been used to study subjects as diverse as wholesale electricity markets (Li and Tesfatsion, 2011; Sun and Tesfatsion, 2007), electricity supplier and smart metering choice (Zhang and Nuttall, 2011), PV-system adoption (Robinson and Rai, 2015; Robinson *et al.*, 2013; Snape, 2016), wood pellet boiler adoption (Sopha *et al.*, 2013), smart grid management (Kremers *et al.*, 2013) and automated demand-side response (Boait *et al.*, 2013), among others. This illustrates the versatility of the technique, which is increasingly recognised within the energy community, with ABM development in energy being recently described in the literature as ‘fertile ground’ (Rai and Henry, 2016). For a detailed view of the ABM methodology, Gilbert (2008) and Miller and Page (2007) are recommended. In short, ABM is a modelling philosophy that focuses on encoding how individual agents within the target system behave and interact with each other, including when they act/interact and how the actions are influenced by conditions. Once simulated agents are coded in software, the system is started, and agents allowed to act. The system behaviour is not defined but can be measured and evolves as a result of the agent actions and interactions. The key benefits of the ABM approach are its ability to model

- heterogeneity in agents – in contrast to standard economic models, ABM can easily accommodate heterogeneous actors in a model
- stochasticity – ABM can be used to model stochasticity in initial conditions; endogenous agent actions; agent interactions; or external conditions

- emergence – as agent behaviours and interactions are defined, overall system behaviour is not, allowing for emergent system behaviours over multiple simulations seeded with different stochasticity
- behavioural models – although perhaps an area for further development, agent behaviours can be based upon established psychological theory rather than intuitive abstractions
- spatial/neighbourhood effects – agents can have a position in the simulated world and therefore to have neighbours that are other spatially proximal agents; this can be on an abstract grid or coupled with a geographic information system (GIS) for realistic placement.

These benefits make ABM particularly suitable for modelling community energy schemes, as the diversity of schemes and localities can be accommodated under a common methodology. The technique allows for heterogeneity within participants in a community energy scheme, both in terms of behaviour and technology ownership as well as being able to incorporate geographic realism and associated effects, for example the effects of localised weather, or of neighbourhood observation and mimicry. Crucially, the method also allows for emergent behaviour of the community, as the constituent agents are defined, but may be seeded with stochastic conditions to test the range of potential outcomes.

With the benefits being outlined, the challenges and disadvantages of the method should also be considered. Often, ABMs are dismissed as 'toy' models, a criticism aimed at either the lack of technical detail and accuracy in agent descriptions, or a lack of scale or the apparently crude abstraction and simplification of behavioural characteristics of agents. Validation and Verification is also an acknowledged difficulty with ABM – as the technique is inherently stochastic, and the number of potential interactions grows exponentially with the number of agents, testing a comprehensive set of model outcomes is infeasible. Thus, ABM is not well suited to modelling where certainty in a singular quantitative outcome is paramount; rather, it excels where accounting for uncertainty and gaining knowledge of the distribution of outcomes and qualitative system behaviour is the aim.

As a consequence, should be stressed that ABM is rarely used in the energy sector for hard, single-valued predictive modelling. Instead, it is useful to investigate the spread of potential outcomes from a policy, theory or scenario and, occasionally, to indicate different modes of operation that may emerge under the same conditions through stochastic variations. Results from ABM simulations should therefore be interpreted with this in mind – as useful insights into possible and probable outcomes, rather than hard monetary predictions. ABM is especially useful in indicating the modes of behaviour that a system will experience under given conditions, rather than specific quantitative predictions of the future.

When combining the advantages with the drawbacks of the techniques, it is apparent that ABM is particularly suited to investigations of certain types. In particular, questions of adoption within a population have proven to be usefully studied via ABM. The ability to combine economic, technological and social factors in the definition of an agent gives ABM a useful role in investigating potential policy impact and theory testing.

8.4. Case studies

In this section, a series of case studies where ABM has been used to provide insight into the potential for community energy schemes are briefly described and analysed. Each case study has been selected to illustrate the utility of ABM in analysing a particular facet of transition to a smart city environment. The studies are developed using the CASCADE ABM framework (Rylatt *et al.*, 2013). Lessons from each are drawn together in the discussion section that follows.

8.4.1 Adoption within households

Significant penetration of community energy schemes will require households to adopt new technology and practices. ABM is ideally placed to combine conventional economic understanding of how adoption decisions are made with psychological theory offering insight into other factors that may affect such decisions. The ability for spatial considerations to be integrated into ABM offer the potential to directly model contagion, where observation of others' adoption can influence a potential adopter's decision. The heterogeneous nature of consumers' predisposition towards adopting various technologies or practice can be accommodated, by quantification of the predisposition on a per-agent basis, by assigning predispositions across a population. A representative spread of predispositions can be either estimated or informed by empirical data from surveys or by using a proxy for the predisposition.

The last approach was undertaken in a model designed to investigate the factors influencing the adoption of rooftop PV systems under the UK Feed in Tariff (FiT) (Snape, 2016; Snape *et al.*, 2016) in order to understand the impact of an incentive policy, in this case the UK FiT. In this model, previous work on a framework for pro-environmental behaviours was used to give an indication of the spread of predispositions towards installing renewable energy, in this case rooftop PV systems. This predisposition was incorporated into a decision-making algorithm based on well-tested psychological theory (the social cognitive theory, SCT), which was used to codify the way in which economic, psychological and social factors were combined to inform the decision-making process within an agent (Figure 8.1). Observation of neighbours was included in the model, with the number of neighbours within an agent's immediate locality being used as a factor influencing the perceived social norm associated with PV-system adoption. The ABM algorithm logic and parameters are shown in Table 8.1 and Figure 8.2.

The model in this case was data informed – it was made spatially explicit by integrating the model with GIS information, the model timestep was aligned to a real-time interval (48 model timesteps = 1 day) and real-world statistics on population distribution were employed alongside the pro-environmental frameworks. The specific community tested in this case was urban and geographically bounded – an illustration of the layout of agents, differentiated by their pro-environmental predisposition category is shown in Figure 8.3. Stochasticity was introduced into the agents' behaviour firstly by allowing the time between agent decisions to vary randomly with a mean of 3 months between

Figure 8.1 Visualisation of the SCT psychological model. (Source: Snape, 2016)

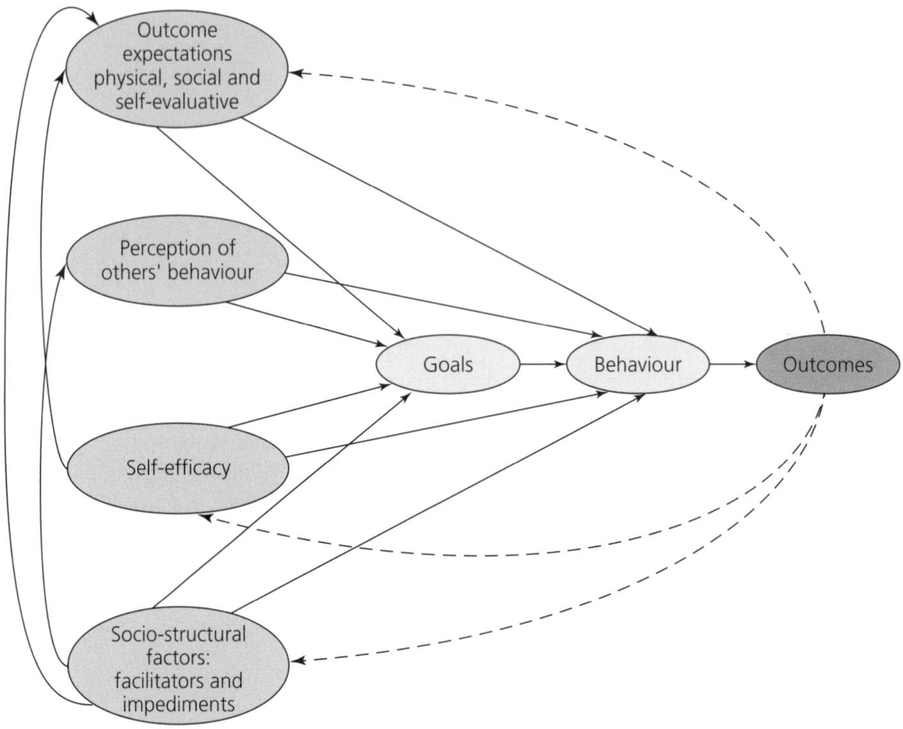

decisions and allowing the agent to have varied radii of observation (meaning that the radius of 'local' for each agent could be adjusted).

Outputs of the model were the penetration of PV systems within the modelled community, the path to get to that penetration (i.e. how that varied between the model start and end) and the effect that had on the day-to-day overall demand of the community. The model showed that neighbourhood observation was a significant factor in the rate and final level of adoption. In common with any stochastic model, many runs should be undertaken and results presented as a distribution of outcomes, rather than a single-valued result. For example, given a particular observation scheme, a single-run adoption pathway will consist of a cumulative number of adoptions at each time-step (Figure 8.4). These may be combined and overlaid to give an indication of the spread of pathways possible (Figure 8.5), and, finally, the distribution of outcomes at a given timestep may be plotted as a histogram (Figure 8.6).

Real-world data showed that, unlike classical adoption curves, the adoption of PV systems exhibited significant discontinuities around the time of FiT policy changes. It was found that by using ABM a useful parameter to include in the model was the anticipation of change in agents – this factor can drive the urgency of decision-making. With this factor incorporated, ABM could model the discontinuities in adoption where more

Table 8.1 Mapping of parameters to SCT constructs

SCT construct	Variables in simulation	Range	Dynamic through simulation or set at initialisation?
Outcome expectations	Expected payback period	+ve integer	Dynamic
Perception of others' behaviour	% neighbours adopting	0–1	Dynamic
	Perception of time until tariff changes	+ve integer	Dynamic
Self-efficacy	Predisposition to green decisions	0–1	Set at initialisation
	Perceived 'hassle' of installation	0–1	Dynamic
Socio-structural factors	Available capital (£)	+ve integer	Set at initialisation
	Building orientation	0–360°	Set at initialisation
	Building heat loss rate (kWh/K/s)	~N(275,75)	Set at initialisation
	Building thermal mass (kWh/K)	~N(12.5,2.5)	Set at initialisation
	Cost of technology (£)	+ve integer	Either, depending on scenario
	Tariff levels (pence/kWh)	+ve integer	Dynamic

Source: Snape (2016).

classical models could not. Interestingly, it was the anticipation of change that drove a large increase in adoption, not a material increase in economic benefit in those short periods. It seems that the fear of missing out on a good deal due to policy change was the driver. This insight can be incorporated into future policy design, underlining the need for stable policy with considered and understood mechanisms to cope with exogenous changes in incentive policy. Such insight demonstrates the potential benefit of ABM for those conducting impact assessments for potential incentive schemes for community energy.

Other studies have used ABM to examine the adoption of sustainable technology by domestic consumers. PV systems have been studied in similar fashion around the world (Islam, 2014; Robinson and Rai, 2015; Robinson *et al.*, 2013), biomass adoption has had a similar treatment (Sopha *et al.*, 2013), and electrical and alternative-fuel vehicles have been a target of such studies (Tran *et al.*, 2012; Zhang *et al.*, 2011). While these studies deal with adoption of a specific piece of technology, perhaps more interesting for community energy purposes is a decision to adopt a different energy supplier, or a whole different business model. There have been some studies towards this end (e.g. Zhang and Nuttall, 2011); however, further work in this regard is needed. Each of these studies differs in the detail of the decision-making algorithm employed, some using models

Figure 8.2 Visualisation of agent decision algorithm. (Source: Snape, 2016)

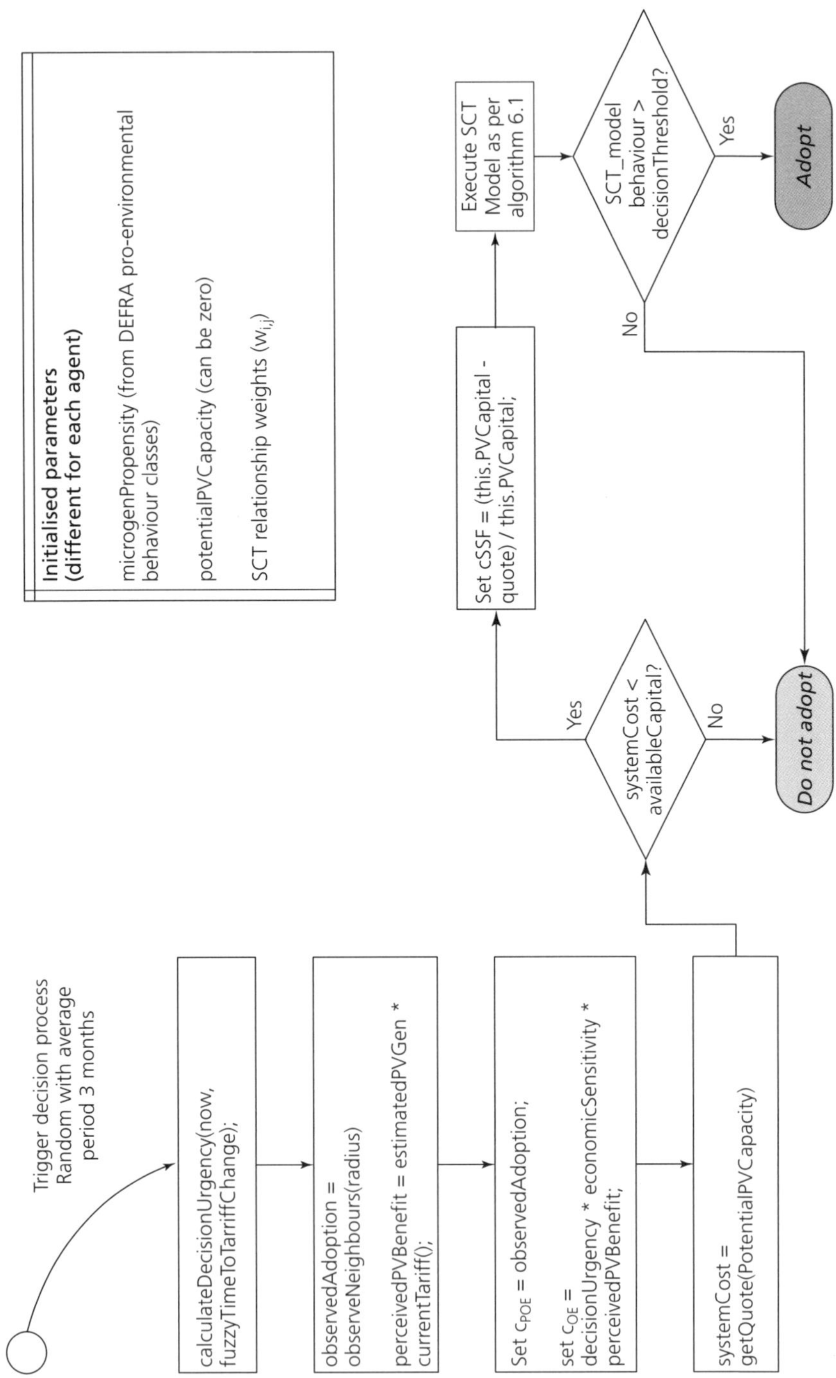

Figure 8.3 Location of a simulated community, postcode LE2 in Leicester, UK. Note the realistic layout of households on streets (far right), enabling detailed testing of neighbourhood observation effects. (Source: ABM described in Snape, 2016)

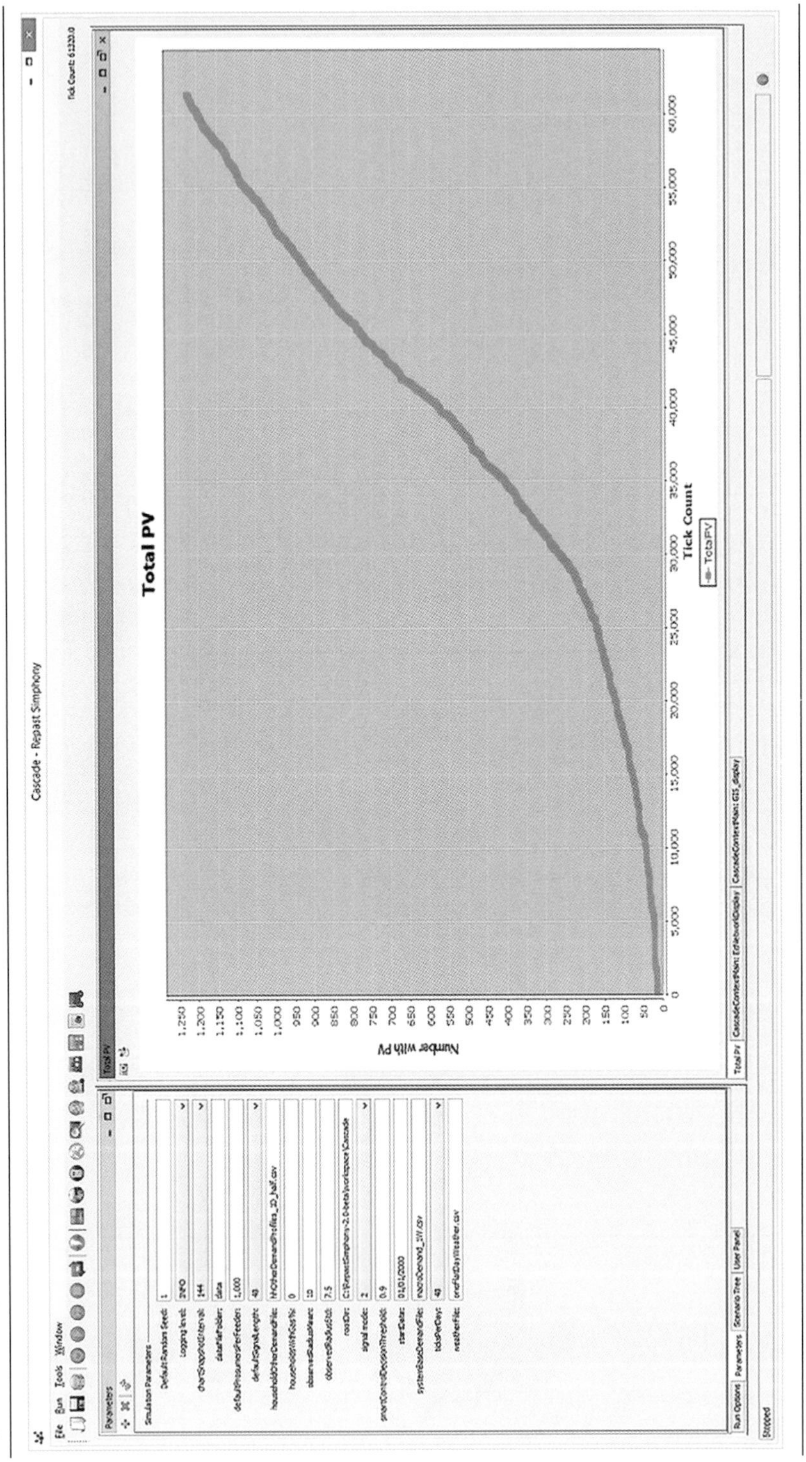

Figure 8.4 Output of a single run of the PV-system adoption model (screenshot)

Figure 8.5 Overlaid adoption pathways from 200 runs

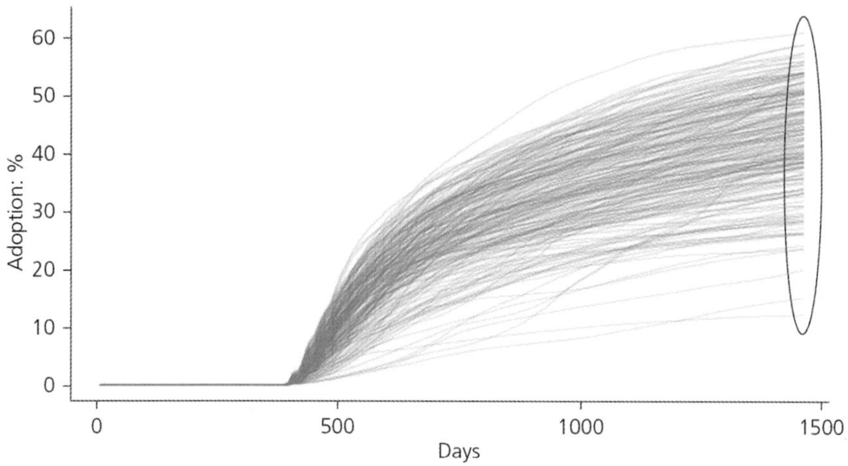

influenced by psychology (the theory of planned behaviour, belief–desire–intention, SCT) whereas others use more intuitively justified decision-making algorithms, or rely on rational economic decision-making, albeit usually with incompletely bounded information.

While these studies differ in the specifics of the decision model encoded and the artefact being adopted, they also have unifying themes and lessons. Firstly, they are driven by a recognition of the inadequacy of classical rational economic models of diffusion and consequent impact in a complex and rapidly changing environment where transition is both being incentivised and occurring. Secondly, they are usually motivated by a desire to represent the heterogeneity and often the geographical distribution of agents adopting

Figure 8.6 Histogram of the final state of adoption within the community after 1500 simulated days. This describes the final states of 200 simulations (indicated by the oval in Figure 8.5)

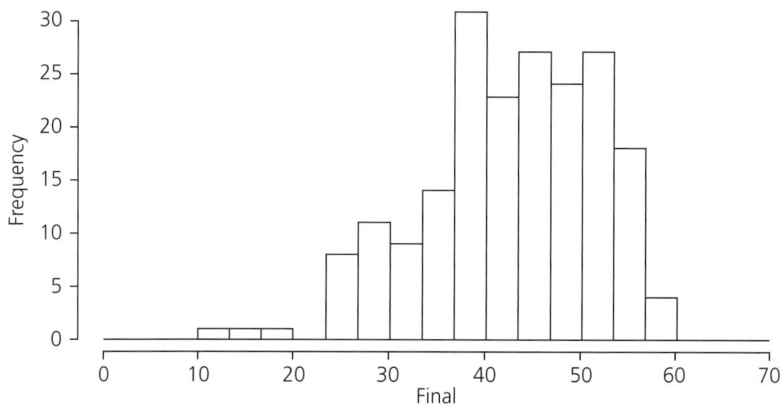

technology or business models. Their results show both the necessity for forecasting models that consider these factors and the potential efficacy of ABM as a tool to create such models.

8.4.2 Automated demand-side response in the developing world

In addition to the behavioural and psychological factors affecting the adoption of renewable energy, technology to manage demand in order to maximise benefit for the community and the individual is often proposed. This technology must first be adopted, but it would also be useful to model the effect of the technology once in use in the community scheme. Here again, ABM can prove to be a useful tool, as individual and heterogeneous demands for each household can be initialised in a model, each agent equipped with a smart controller and the model run to examine how community demand would be changed and the use of renewables improved by their addition. This can be used in a wide variety of community energy schemes, in the UK (Boait *et al.*, 2017) and further afield in the developing world (Boait and Snape, 2014). In this type of ABM, the physics of a dwelling, the particular demands that are needed in the dwelling and the algorithms used in the smart control system can be coded in software, and a wide variety of conditions tested in a 'virtual laboratory' without the need for costly and time-consuming empirical studies in each new community.

To illustrate this case study, a model describing demand management in a very small community in Bhutan is described. In this case, the adoption question was not examined, with all households assumed to have joined the community scheme in question. The issue at hand in this community was primarily one of capacity – a medium-sized hydroelectric generator was the only power source for a community isolated from any larger grid. In this way, the community is representative of many throughout the world and, while not directly related to a city community, allows for a useful case study of a technically constrained community where balancing demand with renewable supply has a direct effect on the quality of service and the security of supply, in contrast to schemes in the UK where the effects on carbon intensity have effects that are remote in time and space to the individual consumers. The ABM modelled a socio-technical issue – a large number of community members wished to use electrical rice cookers to cook rice for a similar eating time. This overloaded the grid, causing a brown-out, which led to a prolonged time with a low quality of supply and long cooking times. A socio-technical solution was proposed, encoding the GridShare scheme described by Quetchenbach *et al.* (2013). The tested smart community solution was to provide households with a device that could distribute the rice cooking in time, while minimising the dissatisfaction felt by consumers (measured as the time in minutes from the ideal time of eating). By using real-time smart signals, the cooking could be slightly distributed (Figure 8.7) to virtually eliminate brown-outs (Figure 8.8) without causing disruption to consumers, without changing any generation or distribution infrastructure.

ABMs of this type can be useful to test the way in which particular smart control algorithms might behave in a variety of contexts to test the robustness of the scheme proposed to different mixes of available renewable generation, appliance ownership, building type and demographic within communities. Crucially, the ABM can also accommodate

Figure 8.7 Distribution of demand before the scheme (day 1) and after its start (day 182). Note the lowered and spread evening peak demand. (Source: Boait and Snape, 2014)

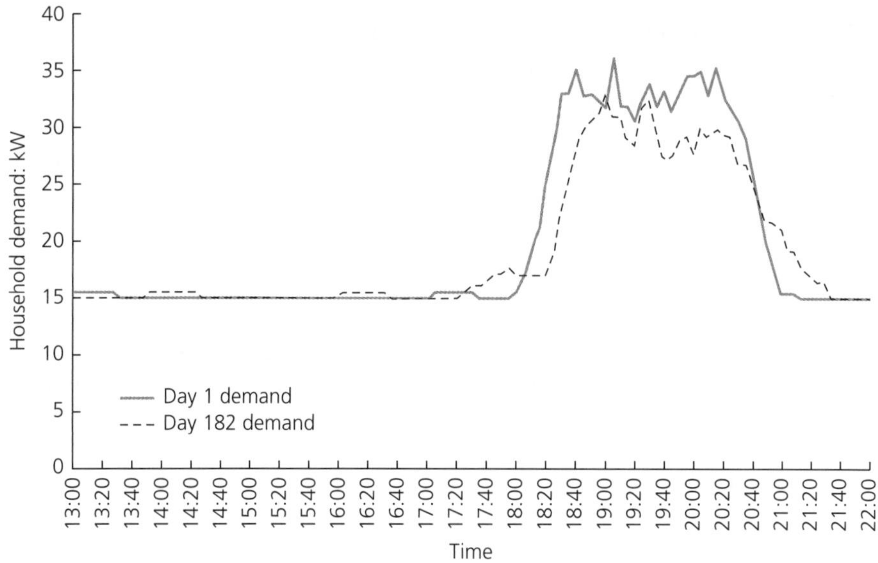

prevalent social conditions, in this case study the desire to cook rice for a given time across the community, but generalisable to any other social conventions of this sort. This is essential to understand how robust a given scheme is to variation in the community where it is deployed. In addition, this model type may be used to examine the potential

Figure 8.8 Evening peak system voltage before the scheme (day 1) and after its start (day 182). Note the much-reduced under-voltage (brown-out) duration and depth. (Source: Boait and Snape, 2014)

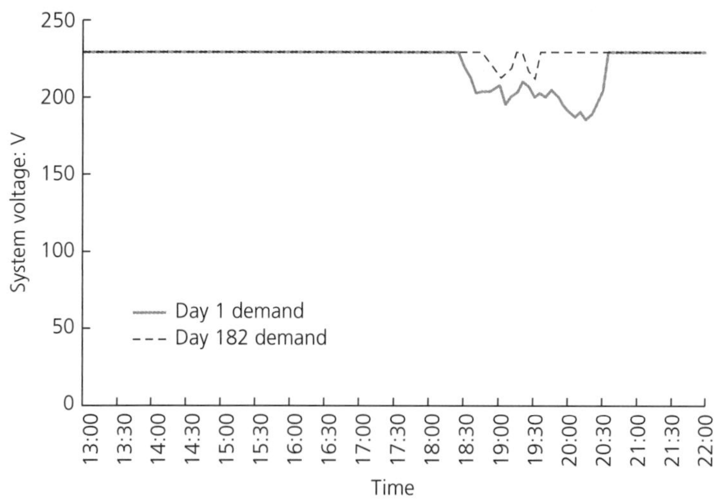

for scaling – testing a given scheme with vastly differing numbers of participants, or with many schemes running simultaneously without the need for real-world trials.

8.4.3 ABM to scale up an integrated community energy scheme

The final case study demonstrates the utility of ABM when investigating the extent to which a community energy scheme might be scaled up by replication across communities with different demographics. The scheme described is the SWELL scheme, which ran a trial community energy scheme across 50 participating households in Shrivenham and Watchfield (villages in Wiltshire, UK).

Briefly, the scheme aimed to incentivise local use of energy generated by PV systems within the participating community. Not all participating households had PV systems, but all could benefit from subsidised electricity prices if using surplus PV energy as it was generated. For regulatory reasons, the subsidy in this trial was paid as vouchers after the fact, rather than altering the supply contract of participating households. Among the participants, a number had dwellings heated by electrical storage heaters, thus integrating the thermal comfort with electricity demand in a number of households. As part of the trial, these households were fitted with smart heating controls. In addition, a number of houses were fitted with domestic batteries, to test the potential for distributed electricity storage within the scheme. Results from the trial were good, with the scheme showing that no participants lost out by participating in it, while some residents gained significantly – particularly those with electrical heating (Boait *et al.*, 2017). The gains in electrical heating were twofold: firstly, simply through smart controls and information from the community energy scheme improving the use of the storage heaters and thus efficiency; secondly, through the smart controller using energy from local PV installations where possible, thus reducing cost.

The next step for the scheme would naturally be to scale up to implement at larger scale, through introducing the model to other communities. This raises a question regarding the potential efficacy of the scheme in different community contexts. ABM proved useful in this regard within the project, and work continues to widen the scope of modelling to investigate the potential for significant penetration of this community energy scheme. In this case, the ABM was firstly encoded with the payment mechanism used by the scheme, essentially a model where PV energy in any half hour was firstly netted off against the owning household. If there was excess across the community after this process, this was shared in a series of rounds among scheme participants who were consuming at the time. Scheme participants were provided with a smart signal, which indicated times when demand was desirable to use PV-generated energy and times to avoid. This was provided via a web portal to any consumer and to the smart controllers of heating and storage where these were fitted.

A particular scenario for the ABM was then constructed to mimic the real-world scheme and run to validate the set-up of the model. To do this, existing models of PV generation along with the smart signal and control regime within the CASCADE framework were used (for details, see Boait *et al.*, 2013). The agents were configured with PV capacities to match those in the scheme, and also electrical heating to match those in the scheme.

A weather file representing the weather experience in the trial was configured, and the model was run. The initial results showed that the actual results observed in the trial fell within the envelope of results output by the model.

The initial correspondence between the empirical and modelled results indicated that the model was useful. A series of tests were then undertaken to investigate how robust the positive results of the empirical trial were to changes in the scenario. Firstly, different weather files were used to examine how dependent the overall results and benefits to individuals were on weather. This showed that the important characteristic of no participant being expected to lose out financially was preserved in differing weather conditions. Next, several communities were programmed to co-exist in the model, and the smart signals sent to consumers in each model were examined after the model had evolved. It was noted that the signals exhibited similar characteristics, but that they evolved in a way that was specific to the community – indicating that the model was transferrable, but that the specific evolution in response to the demographic and technology mix within the community would differ. The signals for each of five 50-member communities are illustrated in Figure 8.9.

Work is ongoing to generalise the model to investigate other interesting scenarios. The performance of the scheme in response to different renewable generation technology (e.g. mini-or micro-hydrogeneration, or community wind) is of interest – particularly as a new empirical trial in a community with micro-hydrogeneration is in progress, giving opportunity for further model validation. Secondly, the integration of heat and mobility with the smart community scheme is of interest – to that end, tests of the same scheme with differing levels of heat pump and EV penetration are planned, with refinement of the models for those technologies under way. Finally, the investigation of how the scheme would work in different sizes of community (e.g. 15 households versus 1500) and differing demographics is of great interest. Intuitively, the initial SWELL community appears to be primed for the adoption of a community scheme, consisting largely of relatively affluent householders who self-selected to participate in the scheme and some of whom had already participated in the purchase of a community-owned wind turbine locally. Plainly, if community energy is to form a widespread part of the more-general electricity system, such demographics cannot be assumed, and the efficacy of the scheme in a less-energy-aware community must be tested. For this task, the ability of ABM to model population heterogeneity could prove invaluable, with the resources and attitudes of the constituent community being encoded into the agents at initialisation and model runs indicating how the scheme may operate in the new scenario. In this way, practical empirical results on a small sample may be extrapolated to the wider population without the need for multiple expensive large-scale trials. However, trials that are conducted may be used to both calibrate and validate the model as results become available.

8.4.4 Renewable incentive-policy design review

This subsection presents an ABM application at a different level of description to the previous case studies. In this case, the ABM model does not represent a case study community scheme, but rather is used to explore the success of renewable incentive policy.

Figure 8.9 Evolved smart signals for the five co-existing community schemes. There is a common shape to encourage demand when PV energy available, but the detailed shape differs according to the community composition (screenshot)

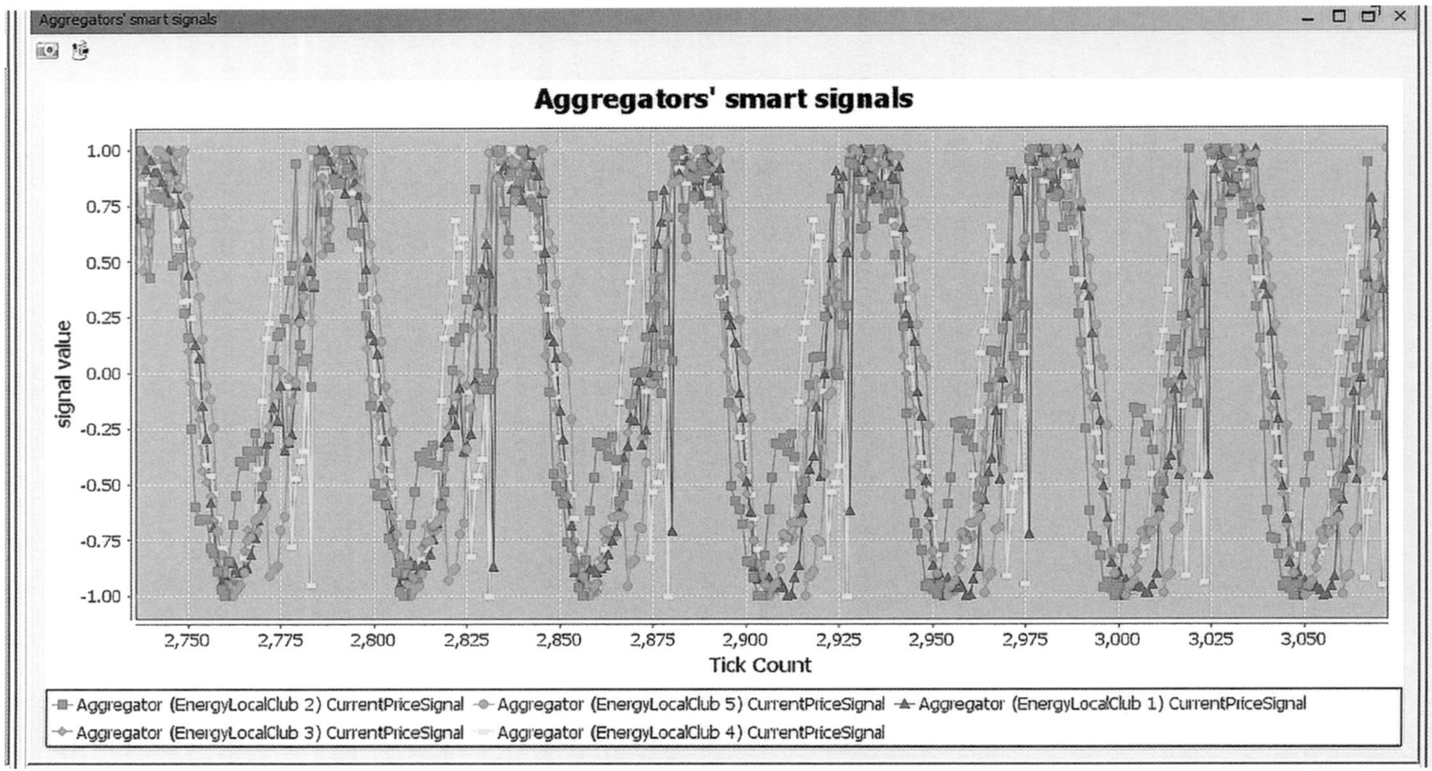

The success or failure of such policies have a direct impact on the success or otherwise of energy schemes within a smart city, motivating the inclusion here. To set the context, the UK currently plans to cut 80% of carbon emissions based on a 1990 baseline (UK Parliament, 2008a) by 2050 with enabling legislation in the Energy Act 2008 (UK Parliament, 2008b). In order to do this, increases in electrically fuelled heating and mobility (EVs) are envisaged, with the increase in EVs and associated infrastructure beginning to become visible on UK roads. The expected contribution of domestic renewable energy technologies (RETs) to this target is outlined in the UK renewable energy roadmap (DECC, 2012). One of the means by which the huge increase in network load implied by such a change may be managed is by the introduction of smart community energy schemes, and these are the focus of both recent policy, practitioner and academic work. Such ideas were considered in Chapter 2.

How might ABM help to analyse and forecast means to move towards this policy goal? One way in which to understand why certain policies are successful is to compare with ones that are less so. Two schemes that incentivise the adoption of domestic-scale RETs, specifically the UK FiT and the Renewable Heat Incentive (RHI) were introduced, with the FiT running 2010–2018 and the RHI in operation. Both policies offer householders who install RETs a payback tariff based on the quantity of renewable energy they can produce.

Broadly, the economic incentive to install the most effective renewable technologies under each policy (PV panels in the case of the FiT and heat pumps in the case of the RHI) was equivalent (Snape *et al.*, 2016); however, the effect of the two policies in terms of their success in stimulating the adoption of domestic RETs was very different.

A previous study using an ABM to investigate the uptake of the RHI alone had found that there was a highly non-linear reaction to non-economic factors, in particular the perceived 'hassle' to the homeowner of installing RETs incentivised under the RHI (Snape *et al.*, 2015). When compared with the FiT, it was demonstrated that while the economic factors were broadly similar for each policy, the non-economic factors in the case of the FiT had a largely positive effect on the rate of adoption, as PV panels are fixed to the exterior of the home, with relatively little disruption to the householder. This makes their installation often very visible – while this occasionally triggers a negative reaction in the observer, a strong effect is the perceived social normality of having PV systems installed. Finally, the changes to policy worsening the economic benefit in an attempt to limit installation actually had the reverse effect, with the consumer downside of 'missing out' on a good incentive triggering extremely high adoption rates immediately prior to policy change. In contrast, for the RHI, non-economic factors have a largely negative effect on uptake. In particular, the hassle of adoption has a large barrier effect. To install a heat pump, at the very least the plumbing of a house must be altered and some sort of heat source (either air or ground based) fitted. Installation of the heat source can be extremely disruptive to the householder, potentially digging up large areas of ground. In addition, between the conception of the policy and its implementation, additional requirements for the efficiency of heating systems were introduced if the householder was to receive the incentive. This often resulted in the need for any adopter to replace all the radiators in the household as part of the installation, further increasing the inconvenience. Finally, post-installation the result is barely visible to neighbours. Thus, the only observation available to neighbours is the rather inconvenient installation process, potentially making visibility a more negative factor in adoption in this case.

While a classical adoption model used in policy-impact assessment would be able to account for these factors only as lumped together externalities, the use of ABM studies of each of the policies and comparison between them was able to highlight these differences and offer explanatory insight into why the FiT was successful to the point of requiring limitation, whereas the RHI failed to meet the adoption rates predicted (Snape *et al.*, 2016).

8.5. Discussion

The case studies presented in this chapter highlight the benefits of ABM for investigating potential futures for city-based community energy schemes. Each study provided results that gave insight into the underlying schemes, sometimes unexpected. If the future electricity system is to include a federation of decentralised community schemes in contrast to the highly formalised hierarchical system of today, a method to quickly analyse policy intended to incentivise such a transition is of great value. Key factors in the transition to a smart energy system providing clean and affordable energy for households in sustainable cities and communities include

- Scale – (how) can successful small-scale schemes scale up?
- Geography – where can community energy schemes be successful?
- Economy – how do the economics of the community context affect potential?
- Policy – how might policies and changes to policies affect the progression towards sustainable community energy schemes?
- Society – how do consumer behaviour and interaction combine to influence the potential for community schemes?

All of these issues are inherently socio-technical, involving people interacting with each other as well as with newly available devices and data. ABM can provide an exciting way to gain insight to all these questions, while incorporating social, psychological, economic and technical factors in agent behaviour, as briefly illustrated in the case studies presented. This begs two questions: What insight can we draw from existing models for community energy schemes of the future? And what future scenario might ABM be used to simulate and test?

The first lesson that we must draw from both the academic literature and modelling is that communities are far from homogeneous, therefore any proposed community energy scheme will give different results depending upon the community where it is launched and the particular configuration of households and technology. This necessitates a modelling method that embraces stochasticity, and we must look at results from models and prospects for smart communities not as a single future but as a distribution of possible futures – a spread of potential scenarios that may occur. The insight that ABM offers here is to show an envelope of likely potential scenarios, with the range of possible outcomes giving an indication of the robustness of the scheme.

From the case studies on PV-system adoption (see Section 8.4.1) and the comparison of FiT and RHI policies in the UK (see Section 8.4.3), we can see that future community energy schemes are heavily dependent on policy drivers and that non-economic factors are crucial in deciding the efficacy of policies designed to incentivise sustainable energy measures in homes. The perceived hassle of installing technology and the observation of others' behaviour are both very strong factors in adoption. Perhaps surprisingly, the strongest driver of all might be a change of policy, as witnessed for the FiT. The lesson is that non-economic factors must be taken into account when designing policy, and ABM can offer a way to test the impact of those factors in combination with economic drivers when policy-makers perform impact assessments. There is more to be done in evaluating the impact of behavioural factors and developing models that are usable for policy-makers during impact assessment. The literature suggests that trust is a crucial factor in the success of community energy schemes – this is an area that remains to be explored via ABM in the community energy arena.

In a smart community energy scheme, the smart components rely on technology. Adoption decisions are crucial here also – and again lessons might be learned from ABM. In addition to response to policy, ABM can also give insight into the technical potential for a scheme. Too often, technologies are tested in a very small number of contexts (sometimes only one) and the results generalised widely by simply multiplying small results for populations en masse. ABM allows for technologies such as a smart controller and the algorithms by which smart signals are sent to be tested in a wide variety of contexts in experiments that can be run cheaply and quickly, as in the case study for the automated demand response described here (see Section 8.4.2) and others in the literature (e.g. Boait *et al.*, 2013, 2017). Unexpected or unwanted economic (Thomas and Tesfatsion, 2018) or technological (Kremers *et al.*, 2013) behaviour can be detected, which corroborates results from other methodologies (e.g. Roscoe and Ault, 2010). Measures to counter unwanted behaviours can be proposed, implemented and tested within the ABM (e.g. Boait *et al.*, 2013). The lesson from ABM here is that it is possible to design schemes for community energy assisted by smart technology that can save money for consumers by better-utilising local renewable generation within the community. This in turn should provide benefits in terms of network capacity, as power flows on larger feeders and substations are reduced. Such benefits are likely to be crucial for the electricity network as reliance on electricity for mobility and heating/cooling increase.

There is much work still to be done. As Rai and Henry (2016) note, 'The real opportunity [for ABM in energy] is at the intersection of theory and applications'. At present, community energy schemes form a tiny proportion of supply arrangements in the UK and worldwide. They are referred to in a number of documents relating to our energy future but are far from ubiquitous in plans for smart cities. ABM can serve to analyse the potential for community energy and integrate economic, social and technical factors to give insight into the range of scenarios possible when introducing community energy schemes in diverse communities. The future for community energy schemes within smart cities is uncertain. On one hand, the importance of sustainable communities and sustainable, affordable energy access is recognised at the highest levels of polity from UN SDGs to the World Bank's energy strategy. On the other hand, in developed countries the path to transition from traditional models of energy delivery to disaggregated community schemes is unclear, while in developing countries the affordability of smart technology for energy is in question. Investigating the feasibility of community energy schemes to unite these goals is of paramount importance, and the case studies presented have shown that

- it is possible to institute community schemes that make use of near-real-time data to both reduce the cost to consumers and lower demands on network infrastructure
- it is possible to institute community schemes that are robust to implementation within differing community demographics, and
- it is essential to consider diversity of population and non-economic factors when designing policy intended to promote adoption of a technology or practice.

These findings are encouraging, along with the demonstration of ABM as a useful tool for enquiry into policy impacts, particularly where adoption of technology or practice is necessary for success and where combining behavioural, technological and economic factors in the decision-making process is important.

There are exciting challenges ahead for community energy. The potential for growth in community energy schemes is evident. Developments in information and communications technology have enabled pioneering schemes to trial platforms enabling peer-to-peer trading of energy within a community (e.g. Energise Barnsley, 2018), and feasibility studies for community demand-side response (e.g. BEIS, 2017) are maturing into small, but fully functional, schemes. Within the smart city, flexibility provided by community energy schemes combined with renewable generation (local or remote) provide a promising way to maximise the utilisation of the existing energy infrastructure to deliver sustainable energy by use of big data from smart devices and meters. ABM can continue to provide a flexible tool with which to model these schemes and to provide insight during a period of rapid transition to smart cities and communities.

8.6. Conclusion

The argument presented in this chapter is that ABM can provide a method to integrate economic factors with the complexity of geography, psychology and social factors to give better insight into how community energy schemes might provide a proportion of the solution to achieving the goals of sustainable communities and access to affordable clean energy in a smart city context. Crucially, ABM provides the means to simulate the scaling up of community energy in a wide variety of environments, obviating some of the cost of pilots and trials.

The value that community energy schemes can bring to a vision for the smart city stems from the increased trust and engagement between customer and supplier that can be engendered by a not-for-profit locally based organisation. This trust and engagement can support consumers and help them to accept, and make effective use of, smart technologies such as automated demand response and business models such as TOD tariffs. The extensive consumer resistance in the UK to the roll out of smart meters by national energy retailers illustrates the difficulty faced by the energy sector in transitioning to a smart city. A complex combination of factors contributes to the outcome of community energy schemes. Economic decisions at the household level are obviously important, but do not give the complete picture. Factors including trust, technology availability and maturity, social networks, perceived social norms and geography contribute to the attractiveness and hence likelihood of success for schemes. Although these effects have been known for some time, policy design, impact assessment and adoption rate prediction are primarily driven by classical economic methods.

Further work is needed. Several areas of investigation where ABM could be useful present themselves. The way in which a group of customers and/or investors is related in order to be termed a community and how energy schemes may flourish within communities present fertile ground for academic enquiry. Often, geographical localisation is understood to define the community, and this definition lends itself to certain types of community energy scheme. However, other communities, for example based on common interest or asset ownership, are possible. Communities within an urban environment may vary from affluent gated developments, through low- and middle-income residents of suburban dwellings to informal settlements where residents may exist on less than a dollar a day. However, from the richest community to the poorest, access to reliable energy infrastructure is a key factor in improving or maintaining quality of life. Exploring how proposed smart community energy schemes can operate within differing models of community presents fertile ground for research.

Investigation of how community energy schemes may work with differing renewable technology mixes complementing or trading-off remains an essential line of enquiry if community energy schemes are to form part of the energy system in the UK and globally. Models of electricity networks and smart grids are relatively more advanced than models of heating/cooling networks and demand. However, the latter represent a large proportion of energy demand, and to implement truly sustainable energy solutions these loads must be addressed sustainably. Current strategy often envisages a synergy of large-scale renewable electricity generation and transition to electrical heating as a path to sustainability. Similar changes are envisaged to decarbonise mobility. This means that models of community energy must increasingly be designed with the integration of electricity, thermal comfort and mobility in mind. This remains an open area for research.

Finally, the place of community energy within other potential paths for transition to sustainable energy provision should be examined. The future of community energy will ultimately depend on the appetite of communities to partake in such supply arrangements. This invites investigation into motivations for involvement in community energy schemes, likely involving models of trust, economics, motivations towards sustainability and technology availability. For those investigations, more lessons may be learnt from future ABM.

REFERENCES

BEIS (Department for Business, Energy & Industrial Strategy) (2017) *BEIS Innovative Domestic DSR Competition – Summary Project Details Phase One*. BEIS, London, UK. https://assets.publishing.service.gov.uk/government/uploads/system/uploads/attachment_data/file/716273/BEIS_Innovative_Domestic_DSR_Competition_-_summary_project_details__1_.pdf (accessed 20/06/2019).

Boait PJ and Snape JR (2014) Demand management for isolated mini-grids supplied by renewable generation. *3rd International Conference on the Developments in Renewable Energy Technology (ICDRET), Dhaka, Banglasdesh*, pp. 1–6.

Boait P, Ardestani BM and Snape JR (2013) Accommodating renewable generation through an aggregator-focused method for inducing demand side response from electricity consumers. *Renewable Power Generation, IET* **7(6)**: 689–699.

Boait PJ, Snape JR, Darby SJ, Hamilton J and Morris RJR (2017) Making legacy thermal storage heating fit for the smart grid. *Energy and Buildings* **138**: 630–640.

DECC (Department for Energy and Climate Change) (2012) *UK Renewable Energy Roadmap: 2012 Update*. DECC, London, UK. https://www.gov.uk/government/publications/uk-renewable-energy-roadmap-update (accessed 20/06/2019).

Dominković DF, Bin Abdul Rashid KA, Romagnoli A *et al.* (2017) Potential of district cooling in hot and humid climates. *Applied Energy* **208**: 49–61.

Energise Barnsley (2018) http://www.energisebarnsley.co.uk (accessed 20/06/2019).

Energy Local (2016) Project SWELL. http://www.energylocal.co.uk/project__swell (accessed 20/06/2019).

Gilbert GN (2008) *Agent-based Models*. Sage, London, UK.

Good N and Mancarella P (2018) Flexibility in multi-energy communities with electrical and thermal storage: a stochastic, robust approach for multi-service demand response. *IEEE Transactions on Smart Grid* **10(1)**: 503–513.

IPCC (Intergovernmental Panel on Climate Change) (2018) *Global Warming of 1.5°C*. IPCC, Geneva, Switzerland. https://www.ipcc.ch/sr15 (accessed 20/06/2019).

Islam T (2014) Household level innovation diffusion model of photo-voltaic (PV) solar cells from stated preference data. *Energy Policy* **65**: 340–350.

Kremers E, González de Durana JM and Barambones O (2013) Emergent synchronisation properties of a refrigerator demand side management system. *Applied Energy* **101**: 709–717.

Li H and Tesfatsion L (2011) Co-learning patterns as emergent market phenomena: an electricity market illustration. *Journal of Economic Behavior & Organization* **82(2)**: 395–419.

Miller JH and Page SE (2007) *Complex Adaptive Systems: An Introduction to Computational Models of Social Life*. Princeton University Press, Princeton, NJ, USA.

Mumford J and Gray D (2010) Consumer engagement in alternative energy – Can the regulators and suppliers be trusted? *Energy Policy* **38(6)**: 2664–2671.

Nerini FF, Tomei J, To LS *et al.* (2018) Mapping synergies and trade-offs between energy and the Sustainable Development Goals. *Nature Energy* **3(1)**: 10.

Pfenninger S, Hawkes A and Keirstead J (2014) Energy systems modeling for twenty-first century energy challenges. *Renewable and Sustainable Energy Reviews* **33**: 74–86.

Quetchenbach TG, Harper MJ, Robinson IV J *et al.* (2013) The GridShare solution: a smart grid approach to improve service provision on a renewable energy mini-grid in Bhutan. *Environmental Research Letters* **8(1)**: 014018.

Rai V and Henry AD (2016) Agent-based modelling of consumer energy choices. *Nature Climate Change* **6**: 556.

Robinson SA and Rai V (2015) Determinants of spatio-temporal patterns of energy technology adoption: an agent-based modeling approach. *Applied Energy* **151**: 273–284.

Robinson SA, Stringer M, Rai V and Tondon A (2013) GIS-integrated agent-based model of residential solar PV diffusion. *Industry Meets Government: Impact on Energy Use & Development. 32nd USAEE/IAEE North American Conference, Anchorage, AK, USA.*

Roscoe AJ and Ault G (2010) Supporting high penetrations of renewable generation via implementation of real-time electricity pricing and demand response. *Renewable Power Generation, IET* **4(4)**: 369–382.

Rylatt M, Gammon R, Boait P *et al.* (2013) CASCADE: an agent based framework for modeling the dynamics of smart electricity systems. *Emergence: Complexity & Organization* **15(2)**.

Snape JR (2016) *Incorporating Human Behaviour in an Agent Based Model of Technology Adoption in the Transition to a Smart Grid.* PhD thesis, De Montfort University, Leicester, UK.

Snape JR, Boait PJ and Rylatt RM (2016) Performance comparison of UK domestic renewable incentives. *Proceedings of the Institution of Civil Engineers – Energy* **169(3)**: 126–139.

Snape JR, Boait PJ and Rylatt RM (2015) Will domestic consumers take up the renewable heat incentive? An analysis of the barriers to heat pump adoption using agent-based modelling. *Energy Policy* **85**: 32–38.

Sopha BM, Klöckner CA and Hertwich EG (2013) Adoption and diffusion of heating systems in Norway: coupling agent-based modeling with empirical research. *Environmental Innovation and Societal Transitions* **8**: 42–61.

Sun J and Tesfatsion L (2007) Dynamic testing of wholesale power market designs: an open-source agent-based framework. *Computational Economics* **30(3)**: 291–327.

Thomas AG and Tesfatsion L (2018) *Braided Cobwebs: Cautionary Tales for Dynamic Retail Pricing in End-to-End Power System.* Department of Economics, Iowa State University, Ames, IA, USA.

Toke D (2005) Community wind power in Europe and in the UK. *Wind Engineering* **29(3)**: 301–308.

Tran M, Banister D, Bishop JDK and McCulloch MD (2012) Simulating early adoption of alternative fuel vehicles for sustainability. *Technological Forecasting and Social Change* **80(5)**: 865–875.

UK Parliament (2008a) Climate Change Act 2008. https://www.legislation.gov.uk/ukpga/2008/27/contents (accessed 20/06/2019).

UK Parliament (2008b) Energy Act 2008. https://www.legislation.gov.uk/ukpga/2008/32/contents (accessed 20/06/2019).

UN (United Nations) (2018) About the Sustainable Development Goals. https://www.un.org/sustainabledevelopment/sustainable-development-goals/ (accessed 20/06/2019).

Walker G, Devine-Wright P, Hunter S, High H and Evans B (2010) Trust and community: exploring the meanings, contexts and dynamics of community renewable energy. *Energy Policy* **38(6)**: 2655–2663.

Zhang T, Gensler S and Garcia R (2011) A study of the diffusion of alternative fuel vehicles: an agent-based modeling approach. *Journal of Product Innovation Management* **28(2)**: 152–168.

Zhang T and Nuttall WJ (2011) Evaluating government's policies on promoting smart metering diffusion in retail electricity markets via agent-based simulation. *Journal of Product Innovation Management* **28(2)**: 169–186.

Part 4. Governance

Energy and Mobility in Smart Cities

ice Publishing

Nuttall, William J, Gibson, David V, Trzmielak, Dariusz and Ibarra-Yunez, Alejandro
ISBN 978-0-7277-6425-6
https://doi.org/10.1680/emsc.64256.167

Chapter 9
Measurement and evaluation of smart city outcomes for smarter governance

Sally P Caird
The Open University, Milton Keynes, UK

9.1. Introduction

Global urbanisation trends are associated with a proliferation of smart city developments enabled by advanced information and communications technologies (ICTs) designed to address contemporary social, economic and environmental challenges and develop innovative solutions across key infrastructures, systems and services in cities and regions. This chapter sets out to examine the relationship between smart city development and smart governance, both the contribution of governance to smart city development and its potential benefits for city governance. Building on an analysis of UK smart city case studies through the SmartDframe research study, this chapter argues that the contribution of smart city developments and their success outcomes to governance is influenced by how cities address the challenges of measurement and evaluation of smart city developments.

In Section 9.2, concepts of smart and smart city developments are introduced, including definitions and types of smart city developments (Section 9.2.1). Next, current work on smart city measurement and evaluation and the associated challenges posed for city governance are outlined (Sections 9.2.2 and 9.2.3). Models of smart city governance are then examined to support understanding of the relationship between governance and smart city development (Section 9.2.4). Section 9.3 introduces five UK smart city case studies included in the SmartDframe study and examines findings to elucidate the relationship between smart city development and smart city governance. The chapter explores how city approaches to the measurement and evaluation of smart city developments contribute to smart urban development and governance.

9.2. A review

9.2.1 Smart city concept

While there are numerous definitions, the British Standards Institution (BSI) defined 'smart' as 'the application of autonomous or semi-autonomous technology systems', claiming that smart city development requires the 'effective integration of physical, digital and human systems in the built environment to deliver a sustainable, prosperous and inclusive future for its citizens' (BSI, 2014a, p. 12). A review of the academic literature suggests that smart cities are described through at least five characteristics (Caird and Hallett, 2018), including

- Integration of ICT infrastructure and technologies and digital transformation of city regions (BSI, 2014b).
- Development of human capital (Hollands, 2008) through ICT-enhanced governance.
- The role of big data, described as information/data assets characterised by high 'volume', 'velocity', 'variety', 'variability' and 'value' for different stakeholders that require high-capacity-data-processing cloud-computing services (Mitchell and Wilson, 2012).
- Differences in smart city development maturity between cities – for example, the International Data Corporation (IDC) (Clarke, 2013) identifies different maturity stages of cities' smart city development.
- The smart city label can apply to a diverse range of newbuild cities and retrofit city developments (Batty *et al.*, 2012). While greenfield smart cities are the exception in the Global North (Shelton *et al.*, 2015), a range of retrofit urban developments are more typical in developed countries. Retrofit smart city developments, include urban development, regeneration and innovation projects (Manville *et al.*, 2014), and may be implemented at different scales across buildings, neighbourhoods, cities and regions (Bosch *et al.*, 2017). The IDC (Clarke, 2013) argues that smarter cities have more smart urban developments of which more are developed at city scale.

Taken together, such smart city characteristics demonstrate the complexity and range of smart urban developments, which raises substantial measurement and evaluation challenges for cities with smart strategies and projects/programmes.

The characterisation of the smart city above can be compared with that introduced in Chapter 1. While there are points of convergence, there are also points of difference. The definition of the smart city remains a contested issue.

9.2.2 Measurement of smart city outcomes

Considerable work is currently ongoing to address challenges associated with smart city measurement and evaluation (see the review by Caird and Hallett, 2018). This includes work on standards around smart city development and performance by the International Standards Organization (ISO), European Committee for Standardization (CEN) and the BSI's PAS 181:2014 'Smart city framework' (BSI, 2014b). There is also significant work driven by the European Commission (EC), including Eurocities CITYKeys (Eurocities, 2019), which provides funding for cities to develop and validate smart city measurement frameworks, measures of key performance indicators (KPIs) and standardised data collection procedures, aiming to enable the comparability, scalability and replicability of smart city solutions.

9.2.2.1 Smart city frameworks and indexes

Currently there is no standardised smart city measurement indicator framework widely accepted by cities to measure performance indicators and evaluate progress against city strategies. The European Innovation Partnership on Smart Cities and Communities (EIP-SCC, 2013, p. 16) observed that 'there is presently no single, broadly-accepted indicator framework that reflects the 'smart city' approach'.

There are many city measurement frameworks and indexes available, although few offer specific smart city measurement indicators and metrics. Some examples include the following.

- The Smart City Reference Model conceptualises the work needed for smart city development through several development stages/layers associated with KPIs, and the corresponding innovation policies required to develop connected, instrumented and integrated infrastructure and related capabilities for intelligent services and innovation (Zygiaris, 2013).
- The Smart City Maturity Model measures the maturity phases of smart city development from city 'ad hoc' project planning, where less smart mature cities are characterised by ad hoc smart projects, to the ultimate 'optimised' city-wide city of systems phase, where more smart mature cities have established smart strategies and more smart city-scale developments (Clarke, 2013).
- The European Smart Cities Ranking (ESCR) model (TU Wien, 2015) measures city outcomes against indicators of 'smart governance', 'economy', 'people', 'living', 'environment' and 'mobility' dimensions (Giffinger *et al.*, 2007).
- The Smart City Index Master Indicators (SCIMI) framework similarly measures 'smart government', 'economy', 'people', 'living', 'environment' and 'mobility' dimensions, although is more focused on digital transformation indicators than the ESCR model (Cohen, 2014).
- The Ericsson Networked Society City Index (Ericsson, 2014) measures cities' ICT maturity against indicators of ICT 'infrastructure', 'readiness' and 'usage' across 'economic', 'social' and 'environmental impact' dimensions (Ericsson, 2014).
- IBM's Smarter City Assessment Tool assesses cities' capabilities in terms of instrumentation, interconnection and intelligence (Dirks *et al.*, 2009).
- The Cities of Opportunity Index of Leading Cities measures 'smart', 'quality of life' and 'economic' indicators (PwC, 2016).
- The CITYKeys indicator framework measures smart city project-level success outcomes and city-level indicators, aligned with EU policies across themes of 'people', 'planet', 'prosperity', 'governance' and 'propagation', with the last described as the potential for upscaling and replication of city solutions (Bosch *et al.*, 2017).

9.2.2.2 Quality of measurement

A detailed examination of available smart city frameworks (Caird and Hallett, 2018) identifies a wide range of measurement approaches, indicators and metrics. Essentially, there is no definitive approach to the measurement of smart city outcomes, although approaches may be appraised against criteria of validity, reliability, credibility and utility, as below.

- **Validity.** Do city measurement frameworks and indicators measure what they claim to measure? To be valid, the selected measurement indicators should be comprehensive and complete without ignoring difficult-to-measure factors, including intangibles, qualitative factors, uncertainties and externalities. Another consideration is the validity of the conceptual model, ideology and values underpinning the measurement framework, and the way complex urban systems are represented.

- **Reliability.** Are city measurement indicators standardised and consistently applied? To be reliable, the selected measurement indicators should permit consistent progress measurement and comparison over time.
- **Credibility.** Are city measurement frameworks and indicators credible to different stakeholders? To be credible, measurement indicators should be trustworthy and believable to stakeholders. Also relevant to credibility is the quality, accuracy, objectivity, comparability, reliability and representativeness of data sources, and the transparency of decisions taken in data handling, normalisation, weighting systems and aggregation methods.
- **Utility.** Are city measurement frameworks and indicators useful? To be useful, measurement indicators should be selected for specific purposes and intended audiences and used appropriately. The usefulness of measures varies, for example, for cities of different sizes, facing different challenges, and engaged with projects developed at different city scales.

The main use of city measurement frameworks has been ranking cities, which Vanolo (2017) criticised as a reductionist technology used to discipline cities. City rankings may provide little insight into the magnitude and quality of differences between cities, although the use of measurement frameworks can enable intercity comparisons and support diagnostics of intra-city challenges. For example, the ESCR model provides smart city profiles of European cities' strengths and challenges (TU Wien, 2015). Such intercity comparisons and intra-city diagnostics can support governance and new policies. Holman (2009) noted that performance measurement indicators can have a useful role as policy instruments in communicating information across networks of policy-makers, government departments and stakeholders, offering the potential to transform governance.

9.2.3 Evaluation of smart city developments

While smart city measurement frameworks are mainly applicable to measuring outcomes at the city level, much work on evaluation takes place at the project/programme level. Evaluation is needed to judge the appropriateness, effectiveness and value of smart city projects/programmes and identify their outcomes for citizens and city regions. The *Mapping Smart Cities in the EU* study (Manville *et al.*, 2014) reports on the evaluation of smart city initiatives and projects based on their goals, scale and scalability, targeted stakeholders, level of citizen engagement and success outcomes. However, evaluation of smart city development programmes has been criticised as non-standard and inadequate, and more focused on implementation processes and investment metrics than on city outcomes and impacts (BIS, 2013).

9.2.3.1 Evaluation process

Edler *et al.* (2012) outlined evaluation processes, including setting evaluation objectives; identifying target groups; articulating key questions and topics; clarifying the proposed city impact dimensions; and setting up the evaluation, including the purpose, timing and commissioning logistics. Spatial and temporal boundaries need to be clarified, recognising that the full impacts of smart city developments may not be knowable for many years (Manville *et al.*, 2014; Tyler *et al.*, 2013). The process also requires selecting methods for data collection, measurement and analysis, and judging the quality, consequences and usefulness of the evaluation (Edler *et al.*, 2012).

9.2.3.2 Evaluation objectives

Evaluation approaches may be informed by theories of change, which describe how the programme objectives should deliver the expected outcomes through a logical framework of linked activities, outputs, outcomes and impacts, with recognition of potential direct/indirect effects and contextual factors (Tyler *et al.*, 2013). This 'theory' (of specific project/programme leaders) could account for targeted/untargeted, intended/unintended outcomes and predictable/unpredictable outcomes that city authorities might identify, as well as the direct/indirect, and beneficial/non-beneficial outcomes that city stakeholders might experience. Evaluation approaches may also be informed by the Forum for the Future's (2007) popular Five Capitals Model, which describes the stocks of capitals (benefits or value) that might be developed, increased or decreased, as a result of new projects/programmes, including 'natural', 'human', 'social', 'manufactured' and 'financial' capitals. Such capitals may be articulated, mapped, measured and tracked against strategic targets, as part of a 'benefit realisation' strategy, a critical success factor for smart city programmes (BSI, 2014b).

9.2.3.3 Evaluation methods

Evaluation questions and the anticipated impacts of a development programme typically determine the choice of methods used for data collection and analysis (Edler *et al.*, 2012). Evaluation approaches may draw on a range of quantitative methods, for example cost–benefit analysis, cost-effectiveness analysis, controlled experiments, before–after assessments, audits, and technical monitoring of performance against baseline measures and strategic KPIs. Available qualitative methods include document analyses, network analysis, citizen surveys, and expert and community stakeholder focus groups (Edler *et al.*, 2012). A range of appraisal and evaluation tools are also available in *The Green Book*, such as design quality indicators, the goals-achievement matrix and multi-criteria analysis (HM Treasury, 2013).

9.2.3.4 Measures, data collection and analytics

Smart city frameworks offer measurement indicators to support the design of evaluation, while work on standards development by the ISO, CEN and BSI is currently establishing standardised smart urban metrics. Moreover, data collection is informed by new data sources enabled by embedded smart ICTs. This includes real-time and near real-time data streams from city infrastructures, sensor networks, contemporary space platform technologies and web services. Data assets also include citizen data – volunteered or crowd sourced – through social media, mobile apps and citizen participation platforms. Smart data assets add to traditional historic/legacy data sources to bring new affordances associated with big data to evaluation processes, including advanced data-processing technologies, predictive analytics, machine learning and stream statistics (Caird and Hallett, 2018).

Different approaches are available to design quality evaluations for smart city developments, building on existing measures, and meeting criteria of validity, reliability, credibility and utility. Drawing on smart city measurement frameworks is helpful for establishing measures applicable to projects/programmes, although it is difficult to establish corresponding measurement indicators that apply to project/programme and city levels of measurement (Bosch *et al.*, 2017). A key evaluation challenge is how to prove the value of smart city developments and impact on city outcomes, by establishing cause–effect relations between interventions and success outcomes in highly complex urban systems. Addressing such challenges is important in view of the consequences of measurement for development systems, and systems of governance and policy actions (Holman, 2009).

9.2.4 Smart governance

Smart governance is an emerging area of scholarship, although complex and difficult to conceptualise. Indeed, it represents the focus of an entire section of this book (Part 4). Smart governance is central to the smart city model of governance integrated with the economy, people, living quality, environment, transport and ICT mobility developed by Fernandez-Anez *et al.* (2017). It is recognised as a critical success factor for smart city development (BSI, 2014b; Raghazi and Finger, 2018). Moreover, it represents an important category of smart city measurement indicators in several city frameworks, including the ESCR model's participation factors (Giffinger *et al.*, 2007), the CITYkeys indicator framework's leadership indicators (Bosch *et al.*, 2017) and the SCIMI framework's digital transformation governance indicators (Cohen, 2014).

9.2.4.1 Conventional and smart governance

Avis (2016, p. 4) defined urban governance as 'the process by which governments (local, regional and national) and stakeholders collectively decide how to plan, finance and manage urban areas'. Razaghi and Finger (2018) argued that conventional governance is driven in three ways: first, by reductionist functional management of urban challenges divided across government silos; second, by the division of decision-making power among key urban actors representing legislative, executive and public administration functions; and, third, by a dominant representative democracy paradigm with relatively short election cycles and time-frames. The first leads to fragmentation of government functions, the second to competition for power, and the third to discontinuity in establishing city visions and agendas.

In contrast, Razaghi and Finger (2018) argued that smart governance can overcome some disadvantages of conventional governance. This was expanded by Fernandez-Anez *et al.* (2017, p. 4) who regarded governance and stakeholder involvement as central to developing smart city strategies across the 'Economy', 'Environment', 'Mobility', 'People' and 'Living' city dimensions. They claimed that new smart governance models are needed to address urban challenges, such as how to achieve greater citizen inclusiveness and participation in democracy, more flexible and resilient models of governance, and greater effectiveness and coordination across government institutions and public bodies responsible for territorial administrations. In their work based on the conceptualisation, implementation and analysis of smart city projects, Fernandez-Anez *et al.*

(2017, p. 5) argued that smart governance models have the potential to transform conventional top-down governance models with new models of citizen participation with government projects to facilitate 'urban co-creation and co-management', and more integrated, multilevel approaches to governance.

9.2.4.2 Dynamics of smart governance

A key dynamic of smart governance is the central role of urban actors, particularly through leadership (Bosch *et al.*, 2017), and city stakeholder and citizen participation in smart city development (Bolivar and Meijer, 2016; Giffinger *et al.*, 2007; Razaghi and Finger, 2018). Fernandez-Anez *et al.* (2017) recognised stakeholder involvement as a key smart governance dynamic, including political actors (e.g. government), knowledge actors (e.g. university), social actors (e.g. citizens and community groups) and economic actors (e.g. industry). Moreover, the framework of Castelnovo *et al.* (2016) emphasises multi-stakeholder participation, placing 'community building and management' for public value generation at the centre of a proposed framework for measuring smart governance.

A second key dynamic of smart governance refers to the technologies enabling digital transformation of city infrastructure, systems and services (Cohen, 2014), including the role of smart ICTs and data utilisation in improving governance and citizen participation. Razaghi and Finger (2018, p. 687) stated that

> This is what can be envisioned as smart governance for smart cities: better governance of urban complexity by utilizing data and a stronger role of citizens in the decision making processes thanks to prevalence of real-time communication and opinion sharing channels.

Different perspectives on smart governance are drawn together in Bolivar and Meijer's (2016) sociotechnical systems model of smart governance, which describes the influence of smart governance strategies and actions, and new governance arrangements on governance outcomes. This includes connected organisational processes, decision-making, e-administration, innovation capacity development, use of smart ICTs, and stakeholder collaboration and citizen participation, as examples of new governance arrangements. This influences three types of smart governance outcomes, including (*a*) transformation of smart governance arrangements, (*b*) changes in relationships between government and other actors, and (*c*) a range of improvements linked to city strategies. Overall, this offers a useful model for examining the role of governance in supporting smart city development through smart city strategies and new governance arrangements, and also exploring how smart city developments and outcomes contribute to smarter governance.

9.3. The SmartDframe study: smart city case studies

9.3.1 Overview

Emerging from this review are questions concerning the relationship between smart city development and governance, both the influence of governance on smart city development and the influence of smart city development outcomes on governance processes.

Figure 9.1 Overview of UK smart city case studies

City	Key initiative
Manchester Population: >545 500 Area: 116 km^2 in North-West England Local government body: Elected mayor working with Manchester City council, a metropolitan borough (part of the Metropolitan County of Greater Manchester)	*Smarter City Programme explores better city working through partnerships and the use of technologies, encouraging investment and development.* ***Key project:*** *Triangulum, CityVerve, Smart airport*
Peterborough Population: >196 000 Area: 343 km^2 in East England Local government body: Elected mayor working with Peterborough City Council, a unitary authority	*'Peterborough DNA', a future city demonstrator programme led by the city council and Opportunity Peterborough, funded by Innovate UK.* ***Key projects:*** *BrainWaves Innovation Portal, Smart Fengate*
Birmingham Population: >1 137 100 Area: 286 km^2 in the West Midlands Local government body: Elected mayor working with Birmingham City council, a metropolitan borough (part of West Midland Metropolitan County)	*Smart City Commission, a cross-sectoral collaboration led by the council and supported by Digital Birmingham. Established Smart City Vision and Roadmap 2013.* ***Key projects:*** *Big Data Corridor, Data Factory Portal, smart traffic control and parking, smart street lighting, REPLICATE SME digital academy*
Bristol Population: >459 300 Area: 110 km^2 in South-West England Local government body: Elected mayor working with Bristol City council, a unitary authority	*'Smart City Bristol', a cross-sectoral and community collaborative programme, led by Bristol Futures and delivered through Connecting Bristol.* ***Key projects:*** *Bristol Future City Demonstrator, Bristol is Open, Bristol Living Lab, intelligent building energy, Bristol Energy Ltd, Venturer cars*
Milton Keynes Population: >255 700 Area: 309 km^2 in South-East England New city (no Royal Charter city status) Local government body: Elected mayor working with Milton Keynes Council, a unitary authority for the Borough of Milton Keynes	*Milton Keynes 'Future City Programme', is associated with 4 UK Catapult innovation centres.* ***Key projects:*** *MK:Smart, MK:Smart Datahub, MK IoT network demonstrator, Electric Bus Trial, LUTZ Pathfinder, UK Autodrive, Our:MK, Community Action Platform for Energy (CAPE)*

A series of case studies were developed, as part of the SmartDframe study linked to the MK:Smart programme (MK:Smart, 2019a), to examine city approaches to smart city measurement, evaluation and reporting in UK cities, including Birmingham, Bristol, Manchester, Milton Keynes and Peterborough (Caird *et al.*, 2016). The cities were selected to represent different-sized conurbations, relevant to their population and geographical area. Case studies were developed during 2015, following interviews with representative city authorities, and documentary analysis of their strategic future city and smart city programmes. Figure 9.1 provides an overview.

Previous work has analysed the case studies to examine city approaches to evaluation and reporting in smart cities (Caird, 2018) and develop recommendations for evaluation design (Caird and Hallett, 2018). This chapter draws on Bolivar and Meijer's (2016) smart governance model to re-examine the SmartDframe case study findings, to explore the role of measurement, evaluation and reporting on smart city development outcomes in mediating the relationship between smart city development and governance.

9.3.2 Smart city vision and strategy

Much of the smart city work undertaken in the cities is influenced by a future city vision of smart connected cities using integrated technologies to address urban challenges and create better places for citizens to live, work, play and travel. The case studies showed that smart city development work was typically embedded in strategic future and smart city programmes and policies led by councils and their directorates in collaboration with the public, private and third sectors, the community and citizen groups (see Figure 9.1).

Bristol City Council established Smart City Bristol, a cross-sectoral and community collaborative programme, led by Bristol Futures, a council directorate, and delivered through Connecting Bristol, the city's digital partnership.

Birmingham City Council established its Smart City Commission, Smart City Vision and Smart City Roadmap in 2013 through partnership with leading figures from business, universities and the public sector, and supported by Digital Birmingham.

Manchester City Council established the Smarter City Programme within the community strategy framework for the future city to explore better city working through partnerships and the use of technologies, encouraging investment and urban development, and delivered through the Manchester Partnership of public, private and third sector organisations.

Milton Keynes Council established the Future City Programme with collaborations between businesses, universities and government partners, including four national Catapult innovation centres.

Peterborough City Council and Opportunity Peterborough – the city's partner economic development company – led city development through their Future Cities Demonstrator programme Peterborough DNA, funded by Innovate UK.

The cities were all in the early stages of implementing smart city development strategies, with the exception of Birmingham, the only city with an established smart city commission, vision and roadmap. However, the smart digital vision was not the only vision guiding the cities. There tended to be a fusion of future visions in cities, based around smart, economically sustainable, resilient, inclusive and liveable cities. For example, Bristol authorities suggested that 'smart cities are about liveable cities' delivering quality-of-life outcomes for citizens enabled by integrated smart technologies. Smart technologies were typically recognised as the means of delivering a strategic future city vision rather than the definition of the vision itself.

9.3.3 Smart governance arrangements

UK cities are characterised by decentralised administrations, which typically require new governance structures established to support work on smart city development. The cities in this study were actively developing as smarter cities, having several funded smart city projects/programmes, and an involvement in both UK and European smart city networks, for example the UK's Core Cities, the Eurocities network of major European cities, and the Small Giants. Several cities, including Manchester and Bristol, have also been working with international partners through the EU–China Smart Cities project to improve public service efficiency through engaging citizens with a co-creation process.

9.3.3.1 Connected organisational processes and decision-making

The case studies showed the cities were beginning to establish more connected organisational processes and decision-making, although the study did not examine e-administration processes (Bolivar and Meijer, 2016). Work on smart city development was associated with cross-sectoral, community and citizen collaborative partnerships. For example, Peterborough City Council organised a smart city leadership event to engage public and private sector organisations in partnership with the Peterborough DNA programme. Moreover, Manchester authorities regarded their leadership in enabling partnership working across sectors as a particular strength: 'The level of collaboration between public, private, academic and the third sector is really strong and powerful, and stimulating transformational change.' There were also new opportunities for government departments to work together with greater integration across government silos. For example, Birmingham and Milton Keynes authorities noted that their smart city work was beginning to transform local government through sharing ideas and data, although services delivery still tended to happen in traditional silos.

The Bolivar and Meijer (2016) model identifies connected organisational processes with key smart governance outcomes. The SmartDframe study showed some evidence of 'changes to government organisation' (Bolivar and Meijer, 2016, p. 687). For example, Birmingham's smart city work was beginning to enable cross-sector partnerships and government departments working together, although this was contested by several city authorities, which criticised a culture of silos and bureaucracy in local government with barriers to data-sharing. Moreover, there was some evidence that smart city work was leading to 'changes in the position of government vis-à-vis other urban actors' (Bolivar and Meijer, 2016, p. 687). This was evident from the range of collaborative partnerships being established, for example in the public, private and third sectors, across citizens and community groups, and through national networks (e.g. the UK's Core Cities) and international networks (e.g. Eurocities, Small Giants and EU–China Smart Cities).

9.3.3.2 Innovation capacity development

Smart city projects may be led by a range of actors, including industry, government bodies and universities. While there were many industry-led smart city projects underway in the cities without government involvement, the councils were leading or partnering on a wide range of smart city projects within the cities. This included areas applicable to governance, economy, education, health, citizen innovation, community participation, living laboratories, environment, data and the internet of things (IoT), and the development of ICT, transport, energy and water infrastructures (see Figure 9.1). The case studies revealed that cities were beginning to transform their capacity for innovation through the development and use of smart ICT infrastructure, stakeholder collaboration and citizen participation (Bolivar and Meijer, 2016) and data utilisation (Razaghi and Finger, 2018) (see Chapter 3).

First, the development of innovation capacity is linked to new smart ICT infrastructure. This is exemplified by Bristol's Future City Demonstrator, which supports digital infrastructure development and the Bristol Living Lab, while Bristol is Open provides an open digital infrastructure to develop the IoT and makes city performance data available through an open data portal, which enables testing solutions relevant to city challenges, such as traffic congestion. Another example is Manchester's Triangulum project, which aimed to transform the Manchester Corridor into a low-carbon smart city district through establishing an electric vehicle transport infrastructure, renovating historical buildings and developing an autonomous energy grid to supply district heating and electricity (Triangulum, 2019). Further examples include the Milton Keynes IoT network demonstrator and Birmingham's smart traffic control and parking and smart street lighting projects.

Second, the development of innovation capacity is linked to cross-sectoral and business participation projects. This is illustrated by Birmingham's Big Data Corridor and Data Factory Portal. Birmingham also offer a small and medium-sized enterprises (SMEs) digital academy programme. Another example is Bristol Energy (2019), set up by Bristol Council as one of the first publicly owned smart sustainable energy companies in the UK. In Manchester, council-led work across collaborative partnerships secured investment for the smart digital and physical infrastructure development around Manchester Airport. In Milton Keynes, the MK:Smart programme led by The Open University with the council and industry works to develop innovative solutions across smart transport, energy, water, enterprise, citizen engagement and education, centred on the MK:Smart Data Hub (MK:Smart, 2019b), which draws together city datasets relevant to city functions. Moreover, Milton Keynes Council worked with industry to establish several smart transport projects, including Milton Keynes' Electric Bus Trial, LUTZ Pathfinder (2019) and UK Autodrive (2019) projects for trialling autonomous vehicles (see also Chapters 1 and 5). A further example is Peterborough's digital platform, which encourages business engagement with the sharing economy focused on the SMEs business area, known as Smart Fengate.

Third, the development of innovation capacity is linked to citizen participation projects. This is illustrated by Birmingham's REPLICATE (Renaissance in Places with Innovative Citizenship and Technology) project, which supports citizen co-creation of innovative smart city services with integrated energy, mobility and ICT solutions. Another example, is Bristol's interesting quality-of-life city dashboard (Open Data Bristol, 2019), which builds on citizen satisfaction data collected across electoral wards. In Manchester, the CityVerve project demonstrates the potential of IoT technologies to engage citizens and communities in healthcare, energy and environment, transport and culture (CityVerve, 2019). In Milton Keynes, the MK:Smart programme has citizen engagement projects, including the MotionMap city-wide transport information service, the Our:MK platform for community engagement, and the Community Action Platform for Energy (CAPE), offering citizens and communities a platform for developing sustainable energy projects etc. A further example is Peterborough's BrainWaves Innovation Portal, which enables citizens to share innovation ideas.

Fourth, cities were developing innovation capacity through data utilisation, enabled by new mechanisms for data generation, collection and sharing, including through data hubs and web-based open-data portals. While not emphasised in Bolivar and Meijer's (2016) model, data utilisation is a strong feature of the work of Razaghi and Finger (2018). Several city authorities recognised that many smart city projects were already delivering significant amounts of data, such as energy, climate change, transport, waste, economic development and liveability data.

The development of data intelligence was beginning to inform city strategies and 'intelligent' interventions. For example, in Bristol, real-time traffic data collected through the Traffic Control Centre with data from the Open Data Portal were analysed to measure and manage city traffic congestion. In Peterborough, city authorities were working on smart tools for data collection and analysis to support appropriate city interventions, for example the ESRI Urban Observatory analytic software (Urban Observatory, 2019), which permits visualisation of complex urban themes such as work, movement, people, public services and systems, and comparison of data from cities.

The case studies showed that smart city projects/programmes were contributing to increasing public availability of open data from UK cities (UK Government, 2019). Bristol city authorities identified the importance of opening up data 'to unlock new opportunities, knowledge and information about the city'. It was recognised that barriers to data-sharing needed to be overcome, including data ownership, privacy and confidentiality. However, a major challenge was deciding what to do with the large quantities of data collected. Some authorities believed that data collection decisions should be guided by a business case, which should be based on whether data would change the challenge context; this would ensure that data collection has value for government and citizens.

9.3.5 Smart city evaluation and reporting

Bolivar and Meijer's model (2016) identifies 'improvements to the city' as a key outcome of smart governance. However, the SmartDframe study showed that demonstrating the success of smart city development projects/programmes is limited by current city approaches to evaluation and reporting, although the cities generally accepted that evaluation was important. For example, Birmingham authorities accepted that 'data was useless without evaluation'. They recognised that 'It is not just about delivering projects, what we want to do is identify where we are making the impacts, how we get things to change, what difference we are making to citizens and businesses.'

9.3.5.1 Limitations of city approaches

City approaches to smart city evaluation and formal reporting were at early stages and mainly driven by external funders' requirements. For example, Bristol authorities recognised that their smart city work needed to contribute clearly to city strategies, although they had not established an evaluation framework to measure the overall impact of their programme. While they favoured a citizen-centric evaluation, they were not clear about the best approach. However, they were concerned that any proposal to add new smart city KPIs might be burdensome, when they already had a statutory obligation to report large numbers of KPIs. Instead, they proposed that 'you just need KPIs about improving the general quality of the city and the contribution made by introducing advanced smart technologies'.

The case studies showed that the cities' evaluation work has been focused at the project level, where councils and their partners typically establish project KPIs covering a range of technical, social, economic and environmental performance measures of progress, as part of the smart city project delivery. The city authorities considered that a good approach was to establish baseline measures against both KPIs and strategic targets to monitor performance and measure progress over time (as recommended by BSI, 2014b; EIP-SCC, 2013). For example, Milton Keynes authorities suggested that 'Without a baseline you cannot benchmark or make any progress. On a project basis or programme basis you must understand what your baseline is, or you cannot say if you have achieved change.' However, they added that many projects are at the early stages, when funders are usually satisfied to see a demonstration of innovation.

Many of the smart city projects were developed as micro-scale innovations, which is why several city authorities were concerned that premature evaluation risked crushing emergent innovation. For example, Manchester authorities were concerned that 'inappropriate evaluation could kill a good idea if conducted too early'. Similarly, Milton Keynes authorities cautioned

> While strategic approaches are required – setting outcomes you want to achieve and measuring whether you are – there are challenges with it . . . but it needs to be flexible and make sure you are adapting to the circumstances of where you are. It would be the kiss of death for some innovation initiatives if you evaluated at the end of 12 months and it was not delivering the outcomes you set in the objectives.

However, they recognised that establishing baseline measurements and monitoring progress over time would help demonstrate the validity of innovation concepts and establish the potential for replication, upscaling, and contribution to city success outcomes.

Plans to evaluate the city-level impacts of smart city developments were also at an early stage. For example, Birmingham's Smart City Commission had progressed work on a city-level evaluation framework, including progress measures aligned with smart city strategies and their Smart City Roadmap. In another example, Manchester City Council worked on an impact assessment framework with several universities for Triangulum. This included a plan to work with stakeholders and monitor how well smart city developments function across transport, energy and utilities infrastructures. Once established, this framework 'can be expanded in scale, geographically and thematically' to apply to city strategies and measures of city performance.

Several city authorities held the opinion that no city could claim yet to have a full city evaluation programme. Peterborough authorities were beginning work on this, and recognised the need

> to have more of a formulated framework that we can assess against, but we also recognise that it will continue to evolve as well, because as we learn more, get more involved and have more access to more data, that influences what we do. We understand that it won't be set in stone.

9.3.5.2 Evaluation challenges

Key evaluation challenges identified by cities centred on choosing suitable methodologies to measure the impacts of smart development work on city outcomes and proving the value for cities and citizens. For example, Birmingham authorities acknowledged there were difficulties attributing causality to smart city activities even when they could demonstrate progress against baseline measures. Similarly, Milton Keynes authorities mentioned they had developed many measures through their MK:Smart programme, which could contribute to a smart city evaluation framework. However, they found it easier to measure project progress indicators than prove complex cause–effect relationships between programmes/projects and city outcomes. Generally, the Milton Keynes authorities preferred to evaluate at the project level, where it is possible to demonstrate that something works, particularly when most of their projects are not ready for city-scale implementation.

9.3.5.3 Developing city approaches to evaluation

The cities were interested in developing city-level evaluation of smart city work linked to city strategies. Some cities were influenced by the BSI's work on its Smart City Framework (BSI, 2014b), and the EC's requirements for funded city projects to provide clear measurement data (e.g. the Eurocities CITYKeys programme; Eurocities, 2019). However, some city authorities criticised smart city frameworks as often based on arbitrary selections of indicators, or focused on easily measurable indicators rather than what should be measured; and either too general or too specific about what a city needed to achieve to become smart.

A synthesis of city authority views suggests the following recommendations. Smart city evaluation should

- be aligned with city strategies, city challenges and circumstances
- be applicable to project, programme or city levels of evaluation
- be appropriate to the innovation development maturity and scale of city developments
- develop through a stakeholder process with citizens and across sectors
- draw on city development and performance measurement indicators and metrics
- include suitable, meaningful, comprehensive measures that reflect the multi-faceted nature of smart cities and complex urban systems
- establish baseline measures against strategic targets and KPIs to measure progress over time and monitor city outcomes
- be flexible, relevant and adaptable to different circumstances and open to improvement and evolution
- build on city data intelligence to support future visions of liveable cities.

While the city authorities recognised that there was no generally accepted approach across cities on how to evaluate smart city development at the city level, some were not convinced of the need for an overarching, standardised smart city framework, which might not be sufficiently flexible, relevant and adaptable to their unique city strategies, challenges and circumstances. Instead, some would prefer smart city evaluation to have a diagnostic purpose, in helping cities to identify the role of advanced smart technologies to address gaps in city development work and innovation opportunities, supporting future city strategies.

9.3.5.4 Reporting on city outcomes

UK cities have statutory obligations and established processes for reporting on measured city performance relevant to city strategies and objectives. This is set out in the community strategies developed by some larger UK cities, or in a city's council plan or corporate plans led by the relevant local government authority. Larger cities also publish annual 'state of the city' reports. However, most city councils have no statutory obligation to formally report on smart city work outside their external funders' requirements. An exception is Birmingham's Smart City Commission, which requires quarterly reports on smart city outcomes for the council's corporate report to the appropriate cabinet member. Another exception is Bristol, where the council's digital partner – Connecting Bristol – produces annual reports on smart city work. Establishing reporting mechanisms applicable to smart city work would help to ensure that smart city work contributes to city performance reporting, policy evaluation and governance.

9.4. Summary and conclusions

This chapter examines the relationship between smart city development and smart governance, using case study analysis to explore how smart city development requires and contributes to smart governance. The study supports understanding that smart governance requires smart city strategies and new governance arrangements to both establish more connected organisational processes and decision-making, and develop innovation capacity through smart ICT infrastructure, cross-sectoral stakeholder collaborations, citizen participation projects (Bolivar and Meijer, 2016) and intelligent data utilisation (Razaghi and Finger, 2018). It also provides support for integrated smart governance models that combine multi-level government approaches with citizen participation and cross-sector collaboration (Fernandez-Anez *et al.*, 2017). The study showed that the transformation of local government organisational structures was beginning in cities, despite continuing barriers to data-sharing and integrated working across government silos.

The case studies, however, provide little information about the contribution of smart city development to smart governance. This is mainly due to limitations of the cities' work on the measurement, evaluation and reporting of smart city developments and their city-level outcomes. Instead, the cities' evaluation approaches were at early stages of development and were typically project focused and funder driven. Without appropriate evaluation mechanisms, there are challenges for proving the value of smart city developments and identifying the impacts on city outcomes, and without established reporting mechanisms there are difficulties in providing feedback on city outcomes to inform governance and policy actions.

This chapter supports claims that there are no widely accepted standardised smart city measurement frameworks (EIP-SCC, 2013), although several cities were working on city-level evaluation of smart city developments. While smart city frameworks/indexes offer useful sources of city-level measurement indicators and metrics (Caird and Hallett, 2018), there was evidence that the cities had a stronger preference for constructing an evaluation approach to demonstrate public value generation and the impacts of smart city developments. This approach should reflect unique city strategies and be flexible, relevant and adaptable to city challenges and circumstances, with an openness to improvement and evolution.

While most smart city frameworks are focused on city-level measurement, the chapter shows that much work on smart city development takes place at project and programme levels, where evaluation is needed. An emerging recommendation is that evaluation should be appropriate to project, programme and city levels of intervention, and to the innovation development maturity and scale of smart city developments. The chapter outlines evaluation processes, including setting objectives to reflect the planned delivery of capitals, value or benefits; designing evaluation in collaboration with key city stakeholders; selecting suitable evaluation methods with valid, reliable, credible and useful measures; establishing baseline measures to measure progress over time against strategic targets and KPIs; establishing data collection methods enabled by smart ICTs and big data analytics; and addressing evaluation consequences through governance and policy actions.

Integrative multi-level approaches to the measurement and evaluation of smart city developments across project, programme and city levels are needed to demonstrate both the impacts on city regions and benefits for citizens, public and private sector stakeholders. Further research is needed to investigate how measurement, evaluation and reporting of the outcomes of smart city projects and programmes can inform smarter governance and policy actions, with attention to capitalising on smart ICTs and big data. In conclusion, while there is support for the importance of smart governance approaches to smart city developments, work on smart city developments has limited potential to inform governance and new policies without addressing challenges of data, measurement, evaluation and reporting.

Acknowledgements

I acknowledge the valuable support of the participating city authorities in Birmingham, Bristol, Manchester, Milton Keynes and Peterborough, and also thank Gerd Kortuem, Stephen Hallett, Geoff Snelson, and Lorraine Hudson for their valuable contributions to the research. The SmartDframe project is funded by the Open University Smart Cities Open Challenge Competition, linked to the MK: Smart programme, which received significant funding from the Higher Education Funding Council for England.

REFERENCES

Avis WR (2016) *Urban Governance: Topic Guide*. Governance and Social Development Resource Centre (GSDRC), University of Birmingham, Birmingham, UK. https://gsdrc.org/wp-content/uploads/2016/11/UrbanGov_GSDRC.pdf (accessed 20/06/2019).

Batty M, Axhausen KW, Giannotti F *et al.* (2012) Smart cities of the future. *The European Physical Journal Special Topics* **214(1)**: 481–518.

BIS (Department for Business, Innovation and Skills) (2013) *Research Paper No. 135. Global Innovators: International Case Studies and Smart Cities*. BIS, London, UK. https://www.gov.uk/government/uploads/system/uploads/attachment_data/file/249397/bis-13-1216-global-innovators-international-smart-cities.pdf (accessed 20/06/2019).

Bolivar MPR and Meijer AJ (2016) Smart governance: using a literature review and empirical analysis to build a research model. *Social Science Computer Review* **34(6)**: 673–692.

Bosch P, Jongeneel S, Rovers V *et al.* (2017) *CITYkeys Indicators for Smart City Projects and Smart Cities*. Eurocities, Brussels, Belgium. http://nws.eurocities.eu/MediaShell/media/CITYkeysD14Indicatorsforsmartcityprojectsandsmartcities.pdf (accessed 20/06/2019).

Bristol Energy (2019) https://www.bristol-energy.co.uk (accessed 20/06/2019).

BSI (British Standards Institution) (2014a) PAS 180:2014. Smart cities terminology. BSI, London, UK.

BSI (2014b) PAS 181:2014. Smart city framework. Guide to establishing strategies for smart cities and communities. BSI, London, UK.

Caird S (2018) City approaches to smart city evaluation and reporting: case studies in the United Kingdom. *Urban Research and Practice* **11(2)**: 159–179.

Caird S and Hallett S (2018) Towards evaluation design for smart city developments. *The Journal of Urban Design* **24(2)**: 188–209.

Caird S, Hudson L and Kortuem G (2016) *A Tale of Evaluation and Reporting in UK Smart Cities*. The Open University, Milton Keynes, UK. http://oro.open.ac.uk/46008 (accessed 20/06/2019).

Castelnovo W, Misuraca G and Savoldelli A (2016) Smart cities governance: the need for a holistic approach to assessing urban participatory policy making. *Social Science Computer Review* **34(6)**: 724–739.

CityVerve (2019) https://cityverve.org.uk (accessed 20/06/2019).

Clarke RY (2013) *Business Strategy: IDC Government Insights' Smart City Maturity Model – Assessment and Action on the Path to Maturity*. International Data Corporation (IDC), Framingham, MA, USA, document #GI240620.

Cohen B (2014) Smart City Index Master Indicators Survey. Smart Cities Council. http://smartcitiescouncil.com/resources/smart-city-index-master-indicators-survey (accessed 20/06/2019).

Dirks S, Keeling M and Dencik J (2009) *How Smart is Your City? Helping Cities Measure Progress*. IBM Institute for Business Value, New York, NY, USA. http://www-01.ibm.com/common/ssi/cgi-bin/ssialias-infotype=PM&subtype=XB&appname=GBSE_GB_TI_USEN&htmlfid=GBE03248USEN&attachment=GBE03248USEN.PDF (accessed 20/06/2019).

Edler J, Berger M, Dinges M and Gok A (2012) The practice of evaluation in innovation policy in Europe. *Research Evaluation* **21(3)**: 167–182.

EIP-SCC (European Innovation Partnership on Smart Cities and Communities) (2013) *Strategic Implementation Plan*. Climate and Energy Fund, Vienna, Austria. https://www.smartcities.at/assets/04-Europa/EIP/Download-Strategischer-Implementierungsplan.pdf (accessed 20/06/2019).

Ericsson (2014) *The Networked Society City Index 2014*. Ericsson AB, Stockholm, Sweden. http://www.ericsson.com/res/docs/2014/networked-society-city-index-2014.pdf (accessed 20/06/2019).

Eurocities (2019) CITYkeys. http://citykeys-project.eu (accessed 20/06/2019).

Fernandez-Anez V, Fernaindez-Gu?ell JM and Giffinger R (2017) Smart city implementation and discourses: an integrated conceptual model. The case of Vienna. *Cities* **78**: 4–16.

Forum for the Future (2007) *The Five Capitals Model – A Framework for Sustainability*. Forum for the Future, London, UK. https://www.forumforthefuture.org/Handlers/Download.ashx-IDMF=8cdb0889-fa4a-4038-9e04-b6aefefe65a9 (accessed 20/06/2019).

Giffinger R, Fertner C, Kramar H *et al.* (2007) *Smart Cities Ranking of European Medium-Sized Cities*. Centre of Regional Science, Vienna, Austria. http://www.smart-cities.eu/download/smart_cities_final_report.pdf (accessed 20/06/2019).

HM Treasury (2013) *The Green Book: Appraisal and Evaluation in Central Government*. HM Treasury, London, UK. https://www.gov.uk/government/publications/the-green-book-appraisal-and-evaluation-in-central-governent (accessed 20/06/2019).

Hollands R (2008) Will the real smart city please stand up? *City* **12(3)**: 303–320.

Holman N (2009) Incorporating local sustainability indicators into structures of local governance: a review of the literature. *Local Environment* **14(4)**: 365–375.

LUTZ Pathfinder (2019) https://ori.ox.ac.uk/projects/lutz-self-driving-pods (accessed 20/06/2019).

Manville C, Cochrane G, Cave J *et al.* (2014) *Mapping Smart Cities in the EU*. Directorate-General for Internal Policies, Policy Department A: Economic and Scientific Policy, European Union, Brussels, Belgium. IP/A/ITRE/ST/2013–02, PE 507.480. http://www.smartcities.at/assets/Publikationen/Weitere-Publikationen-zum-Thema/mappingsmartcities.pdf (accessed 20/06/2019).

Mitchell I and Wilson M (2012) *White Paper: Linked data: Connecting and Exploiting Big Data*. Fujitsu Services, London, UK. https://www.fujitsu.com/uk/Images/Linked-data-connecting-and-exploiting-big-data-%28v1.0%29.pdf (accessed 20/06/2019).

MK:Smart (2019a) http://www.mksmart.org (accessed 20/06/2019).

MK:Smart (2019b) MK:Data Hub. https://datahub.mksmart.org (accessed 20/06/2019).

Open Data Bristol (2019) Quality of life in Bristol 2017–2018. https://opendata.bristol.gov.uk/pages/quality_of_life_results_201718 (accessed 20/06/2019).

PwC (2016) *Cities of Opportunity*. PwC, London, UK. https://www.pwc.com/us/en/library/cities-of-opportunity.html (accessed 20/06/2019).

Razaghi M and Finger M (2018) Smart governance for smart cities. *Proceedings of the IEEE* **106(4)**: 680–689.

Shelton T, Zook M and Wiig A (2015) The actually existing smart city. *Cambridge Journal of Regions, Economy and Society* **8(1)**: 13–25.

Triangulum (2019) http://www.triangulum-project.eu (accessed 20/06/2019).

TU Wien (Vienna University of Technology) (2015) European Smart Cities 4.0. TU Wien, Vienna, Austria. http://www.smart-cities.eu/index.php-cid=7&ver=4 (accessed 20/06/2019).

Tyler P, Warnock C, Provins A and Lanz B (2013) Valuing the benefits of urban regeneration. *Urban Studies* **50(1)**: 169–190.

UK Autodrive (2019) http://www.ukautodrive.com (accessed 20/06/2019).

UK Government (2019) Find open data. https://data.gov.uk (accessed 20/06/2019).

Urban Observatory (2019) http://www.urbanobservatory.org (accessed 20/06/2019).

Vanolo A (2014) Smartmentality: the smart city as disciplinary strategy. *Urban Studies* **51(5)**: 883–898.

Zygiaris S (2013) Smart city reference model: assisting planners to conceptualize the building of smart city innovation ecosystems. *Journal of the Knowledge Economy* **4(2)**: 217–231.

ice Publishing

Nuttall, William J, Gibson, David V, Trzmielak, Dariusz and Ibarra-Yunez, Alejandro
ISBN 978-0-7277-6425-6
https://doi.org/10.1680/emsc.64256.187

Chapter 10
Energy companies in transition: seeking legitimacy or legitimation?

Tahrir Jaber and Elin M Oftedal
The Arctic University of Norway

10.1. Introduction

Energy plays an essential role in cities, as most everyday activities involve the use of energy sources for light, transport and heat (Calvillo *et al.*, 2013). However, while the traditional city may use electricity generated from coal-fired power stations, or fossil fuels, in the smart city concept, energy should both 'meet the needs of the city and its citizens' and be 'clean and sustainable' (Calvillo *et al.*, 2013; Pellicer *et al.*, 2013). Sustainable development (SD) of the smart city is planned through different projects such as sustainable urban infrastructure, intelligent transport systems and smart mobility, and through the development of renewable energy (RE) (Pellicer *et al.*, 2013). Many scholars and policy-makers envision the 'smart city' as becoming the engine of SD (Batagan, 2011; European Commission, 2016).

SD emphasises the long-term perspective in terms of economic and social development, which enhances the natural environment and social equity (Diesendorf, 2000; Elgazzar and El-Gazzar, 2017). Thus, to achieve sustainability, smart city initiatives require society – citizens, government, business and non-government institutions – to make a change (Albino *et al.*, 2015; Vanolo, 2014).

How to create sustainable cities and smart cities through strategies and development projects has been widely discussed in the literature (Giffinger *et al.*, 2007; Vallance *et al.*, 2012). Companies represent smart cities as outcomes or business opportunities (Woetzel and Jaana, 2018). In addition, a smart city provides innovative solutions, better public services and better use of resources, it also, as noted in Chapter 9, guides how people govern and make sustainable decisions to ensure an SD future (Ludlow, 2016). However, the evolution of sustainable smart cities remains under-studied and needs further conceptual and empirical research (Kitchin, 2015). Therefore, the previous chapter presented different smart city projects in the UK and explained how these projects are led by several actors, including industry, government, the city and universities. In this chapter, however, we focus on the role of government towards a 'green shift' and how an oil and gas (OG) company develops sustainable smart cities.

According to Yigitcanlar *et al.* (2018), smart 'communities' are seen as the most essential driver of smart city development. Community engagement projects contribute to local economic development and SD by, for instance, providing access to new technologies and services, and by seeking to modify the behaviours of local communities (Hughes and Spray, 2002; Yigitcanlar *et al.*, 2018). Furthermore, OG companies have placed a significant amount of investment in different types of RE projects (solar, wind, geothermal power, etc.) (Csomós, 2014).

To increase our understanding of and more critically evaluate sustainability approaches for building smart cities, studying organisational shift as an a priori necessity may give clues towards a theory. This chapter deals with a well-established and strategically important OG company's move towards SD to become a broad energy company, with a special focus on renewable sources.

The OG company Equinor (formerly Statoil) is owned mostly by the Norwegian government. This government has committed itself to a 'green shift' – in other words towards SD – at all levels of society. This also involves companies it owns. Therefore, we find it interesting to understand how government policy is legitimated within Equinor.

However, companies enact and uphold the socio-cultural norms in their institutional environments (Handelman and Arnold, 1999; Palazzo and Scherer, 2006). Government-led or popular pressure for a 'green shift' creates particular difficulties for OG companies. Indeed, OG companies face a number of paradoxes: on the one hand, for instance, OG companies are legitimised through profitability from the delivery of a public good (energy); on the other hand, negative externalities from their operation lead to a lack of public acceptance. At the same time, key stakeholders – not least shareholders – may be sceptical of the introduction of RE sources, as they are viewed to be less profitable and effective.

Consequently, OG companies need to establish legitimacy for their new strategy (Du *et al.*, 2007; Yoon *et al.*, 2006). Several studies of environmentally sustainable behaviours provide important insights by focusing on companies as the unit of analysis (Czinkota *et al.*, 2014; Thomas and Lamm, 2012), but what role legitimacy plays in developing a new sustainable strategy has rarely been examined (Thomas and Lamm, 2012). By using legitimacy as a unit of analysis, this chapter aims better to investigate the following research question: 'What is the role of legitimacy in creating a shift towards SD?' The chapter will thereby explore and seek to explain how an OG company strengthens its commitment to the environment and invests in RE activities in order to develop an all-encompassing energy company.

Responding to the request for more qualitative studies on the micro-level of legitimacy in relation to sustainable practices (Binz *et al.*, 2016; Kishna *et al.*, 2017), a single case study from the OG sector was chosen because of its specific challenges with balancing 'standard' activities with demands for improved sustainability (economic, social and environment factors). By using semi-structured interviews, it was possible to delve deeper into the legitimacy process, where three main aspects of legitimacy were compared (i.e. moral, pragmatic and cognitive).

This chapter reframes governance as an opportunity for stakeholders to engage in a sustainability decision-making process, by exploring how key internal stakeholders in Equinor view the 'green shift', using internal legitimacy as a conceptual tool to understand this process. The remainder of this chapter is organised as follows. First, a review of the literature relevant to sustainability and legitimacy is provided. Second, a research methodology is described. Third, exploring Equinor as a single case study, the findings of this study are presented and three propositions regarding new sustainable transition practices are derived. Finally, the implications and limitations of this study, and directions for future research, are considered.

10.2. Context: sustainability and legitimacy

10.2.1 Rationale for SD in an OG company

The Brundtland Report was the first report that looked at sustainability and environment in detail and defined SD as 'development that meets the needs of the present without compromising the ability of future generations to meet their own needs' (UN, 1987). Increasing sustainability is considered to be a socially, environmentally and economically responsible (Taherdangkoo *et al.*, 2017), and SD is increasingly considered part of a company's long-term strategy to improve its social, environmental and economic performance (Timothy *et al.*, 2015) with a shift towards corporate sustainability containing various operating practices (Kotler and Lee, 2005).

Meanwhile, an expanding research stream examines sustainability as a marketing strategy (Azmat and Samaratunge, 2009) or a tool for a competitive advantage (Taherdangkoo *et al.*, 2017), where companies engage in social and environmental activities to seek society's acceptance (Kumar *et al.*, 2012). These studies gave rise to accusations of 'greenwashing', where sustainability was sought to make the business look good but where substantive steps towards a more sustainable operation were avoided.

Such research has provided important insights into why sustainability is sought from a business point of view; however, they may not be fully able to explain *how* companies develop sustainability. Fisher and Bonn (2011) argued that sustainability is seen as a missing ingredient in corporate strategy that needs to be considered in strategic decision-making. In addition, Engert *et al.* (2016) provided a general framework concerning the integration of corporate sustainability into strategic management, where investment, clean innovation, internal commitment to social and environmental responsibility and competitive advantage are substantial steps towards SD.

A large share of sustainability studies on corporations has been based on structures, techniques and learning modes to enhance sustainability in managerial and employee practices (Frandsen *et al.*, 2013). In 2001, Epstein and Roy emphasised that managing corporate sustainability formulation and its implementation requires the ability to translate strategy into action. This translation is supported by the development of appropriate plans, programmes, systems and performance indicators in order to achieve sustainability goals and objectives (Epstein and Roy, 2001). Fifteen years later, however, Engert *et al.* (2016) suggested that there is a lack of empirical studies on the integration of corporate sustainability and how it is carried out in practice – such as through internal communication or manager attitude and behaviour.

Thus, according to Taherdangkoo *et al.* (2017), there is a need for more in-depth research to understand the nature of adopting sustainable strategies.

10.2.2 Legitimacy and sustainable practices

The ability to attain stakeholders' judgement of appropriateness is central to the survival and success of firms (Guerreiro, 2015; Lamin and Zaheer, 2012; Petkova, 2016). Established firms usually have high legitimacy; however, when circumstances push them to change, these firms need resources and experiences from both internal and external stakeholders to create an initial base of legitimacy (Bitektine, 2011; Tolbert and Zucker, 1983; Zimmerman and Zeitz, 2002). In addition, it is more difficult for firms to gain legitimacy in a smart city due to the government roadmaps for improving the quality of citizens' life in terms of energy supplies, transportation system, education, health and innovation (England, 2009; Lee *et al.*, 2013).

Until recently, OG companies have seemingly felt confident claiming that they gain legitimacy through delivering goods and services to the marketplace, providing jobs, treating their employees well, rewarding investors with profit, complying with legislation and paying taxes. Following this logic, there is, therefore, no need to enhance their legitimacy through investment in sustainable activities. However, researchers have argued that the OG sector has increasingly experienced difficult reputational damage and legitimacy crises because of their damage to the environment when producing OG (Freudenburg and Gramling, 2011; Sarker, 2013). Thus, global environmental challenges and the public pressure on OG companies to evaluate their activities and emphasise the importance of legitimacy (Idowu *et al.*, 2013) has led some OG companies to rethink their downstream strategies, with the aim of diminishing their carbon footprint and developing alternative, cleaner sources of energy such as RE (Csomós, 2014; Wan Ahmad *et al.*, 2016).

In order to manage strategic change, address external concerns and increase wider public legitimacy, legitimacy must also be attained internally (Bridwell-Mitchell and Mezias, 2012; Clark *et al.*, 2010; Gioia *et al.*, 2000). Securing legitimacy from internal stakeholders is essential during organisational change in order to align organisational identity with the desired future image (Bridwell-Mitchell and Mezias, 2012; Clark *et al.*, 2010; Gioia *et al.*, 2000).

More than this, however, according to Deegan (2014), legitimacy is used to provide directions about what organisations should do. Suchman (1995) defined legitimacy as 'a generalised perception or assumption that the actions of an entity are desirable, proper, or appropriate within some socially constructed system of norms, values, beliefs, and definitions'.

In order to understand the SD process in Equinor, we considered three analytical elements of legitimacy as defined by Suchman (1995): moral, pragmatic and cognitive.

Moral legitimacy includes evaluation of a company and its new business goals (Aldrich and Fiol, 1994; Parsons, 1960), and determines whether the new business activity is 'the right thing to do' not only for the company's benefit, but also, through a 'social lens', in order to see if the action is widely accepted (Bloodgood *et al.*, 2017). Moral legitimacy is thereby aligned with current social mores.

Pragmatic legitimacy indicates the self-interested calculations of a company's direct stakeholders (Bloodgood *et al.*, 2017). In other words, direct stakeholders estimate a value and choose a particular set of constituents enabling a new venture (Bloodgood *et al.*, 2017; Dowling and Pfeffer, 1975). This pragmatic legitimacy can differ from public opinion and be aligned to the interests of certain groups of stakeholders. In this example, pragmatic legitimacy would be given to what is considered good business opportunity (Bloodgood *et al.*, 2017).

Unlike moral and pragmatic legitimacy, cognitive legitimacy is based on cognition rather than evaluation or interest (Aldrich and Fiol, 1994; Scott, 2014). It involves either affirmative support for a company or an enhanced acceptance of a company based on some taken-for-granted cultural account (Bloodgood *et al.*, 2017). Cognitive legitimacy involves knowledge or product acceptance that a company uses or produces, and considered as a state of the art, where the highest level of cognitive legitimacy is 'taken-for-grantedness' (Bitektine, 2011). As Jepperson (1991, p. 147) noted, a 'taken-for-grantedness' can involve positive, negative or no evaluation.

The empirical study here seeks to consider the three types of legitimacy expressed at Equinor and their facilitation of the company's transition towards SD. In doing so, the chapter explores the complex set of actors and resources involved in the development of sustainable practices in an OG company.

10.3. Research method

A longitudinal single case study design was used in this study to 'explore a real-life contemporary' case (Creswell, 2012). While single case studies are sometimes seen as having limited research value, this is, according to Flyvbjerg (2006), due to some common misconceptions. However, Flyvbjerg (2006) and Jacobsen *et al.* (2002) argued that a single case study can be valuable, not least as it gives scope for the production of 'context-dependent knowledge' over an extended period of time (Eisenhardt and Graebner, 2007), enabling researchers better to understand theoretical constructions (Dyer *et al.*, 1991), as well as providing rich sources of data (Yin, 2014b). Indeed, Beveridge (1951, quoted in Flyvbjerg, 2006) claimed, 'more discoveries have arisen from intense observation than from statistics applied to large groups'.

Thus, the case study method allowed us to use varied and rich sources of data in order to understand the nature of legitimacy that moved Equinor towards an SD transition. This also enabled us to understand interviewees' different viewpoints and personal explanations.

10.3.1 Selection of empirical settings

The Norwegian context is interesting due to Norway's commitment to the UN's 17 Sustainable Development Goals and the Norwegian regulations that prioritise sustainability

(Ødegaard and Grønning, 2017). Since 1987, Norway has made legislative changes to reduce the climate change effects of greenhouse gases, and is working towards an ambitious global climate agreement to ensure environmental sustainability (Brundtland, 1987; KLD, 2014). This made for significant changes regarding environmental policy, where RE has been introduced as an alternative clean energy and is promoted as a climate change adaptation. At the same time, however, Norway's petroleum activities are crucial for the country's economic growth, and for funding the Norwegian welfare state. In addition, like other nations, Norway is contending with contrasting needs for more energy and commitments to a low-carbon future.

Equinor (formerly Statoil) is an international energy company that is mostly owned by the Norwegian state. According to Equinor's own reports, since the 1990s it has worked on achieving its sustainable business by reducing its carbon dioxide emissions produced by its OG operations. Equinor also claims to have strengthened its commitment to sustainability by offering alternative, clean energy at a lower cost, particularly since 2014, when Eldar Sætre started as Equinor's new CEO. Equinor claims to be committed to providing the world's energy needs in a responsible manner, and to creating innovative business solutions. This poses challenges for Equinor to be part of the low-carbon transformation and to fulfil the company purpose of turning natural resources into energy. Through the research period, the organisation has gone through a major shift in focus from solely OG to becoming a mixed-energy-resource company, including RE. This marked a fundamental change for the company and precipitated a rebrand and name change, from Statoil to Equinor.

Between 2014 and 2017, Equinor invested 400 million NOK (the currency of Norway) to develop prototype offshore floating wind turbines. The plan was elaborated in 2008 (Equinor, 2008). Despite the significant investment in this RE platform, according to one interviewee the public believed that this was not a genuine transformation, but instead was intended merely to enhance Statoil's reputation and gain better market access in Northern Norway.

Large-scale commercial offshore wind investment took place in 2012, when Statoil invested £1 billion in Sheringham Shoal (UK) (Equinor, 2019a). However, more substantive movement towards sustainable energy developed a few years later when Statoil invested heavily in RE projects in a variety of locations: Dogger Bank (UK, 2016), Dudgeon (UK, 2017), the Hywind Scotland Pilot Park (UK, 2017), Arkona (Germany, 2019) and Empire Wind (USA, estimated first power in 2025) (Equinor, 2019b).

Such a complex context suggests that using Equinor as a case study on the nature of legitimacy and sustainability in an energy company could well be fruitful.

10.3.2 Data collection

Interviews with Equinor employees collected over a 3-year period (2014–2017) followed a semi-structured narrative approach (Yin, 2014a). We based the interview guide on the three dimensions of legitimacy and SD and the nature of Equinor's transition towards

SD. For instance, to probe moral legitimacy, questions such as 'Do you think SD transition is the right decision for Statoil?' and 'How would you describe Statoil interest in SD transition?' were raised. For pragmatic legitimacy, we asked questions such as 'What kind of competence (employee background, managerial competence/charisma) do you have to support SD transition?' While for cognitive legitimacy, we sought to open up a 'chain of thoughts', with probes such as 'How did you find working in renewable energy projects and making decisions compared with OG?' and 'How do renewable energy projects influence Statoil's image?' In accordance with the respondents, we interpreted SD to mean either investment in new clean energy projects such as RE *or* improved OG processes.

Most interviews took place in Equinor's office in Oslo, with some interviews being conducted by video call (Skype). We selected 15 key persons to interview, including the following roles: senior consultant on sustainability, project manager, research and development engineer, strategy innovation manager, head of innovation special projects, head of corporate strategy development, former chief economist, responsible for wind operation, responsible for carbon capture and storage, senior vice president, head of wind assets management, special consultant for exploration activities, vice president and principal engineer in the mapping department.

Secondary data were collected through Equinor's own documentation, such as company webpages, annual reports, sustainability reports, RE reports, energy perspectives reports, conferences and presentations. Other secondary data sources included news articles and interviews with Equinor staff on local TV/radio.

10.3.3 Data analysis and verification

We used direct content analysis to code our primary and secondary data (Zhang and Wildemuth, 2009), summarising the raw data into categories or themes (Patton, 2002).

Thus, the initial broad coding started with a deductive analysis based on legitimacy theory. For moral legitimacy, we looked for statements indicating an acceptance of the companýs responsibility to 'do the right thing'. For pragmatic legitimacy, we looked for an understanding of the SD transition based on the self-interest of the organisation. Finally, for cognitive legitimacy, we looked for an endorsement of the knowledge needed for the SD transition.

For the second stage of data analysis, following Zhang and Wildemuth (2009) and Hsieh and Shannon (2005), we used an inductive approach that involved identifying coding categories derived directly from the raw data in a 'line by line approach', where we looked at each sentence separately and allowed themes to emerge.

A pattern-matching technique was used to compare the patterns of results obtained from our study with patterns from past studies, knowledge and theory (Gibbs, 2002; Yin, 1994). Thus, the pattern-matching technique was used to examine the correctness of the propositions, and build explanations on whether and why the patterns matched or not (Almutairi *et al.*, 2013).

A data reduction process was followed to enable better comparison of the different types of data (Miles *et al.*, 1994), with coding based on 496 references from both primary and secondary data. As the analysis proceeded, the main logical frame shifted from exploring data using retroduction (refine and redevelop theory) to verifying theory through deduction (develop propositions and test them) (Van De Ven and Poole, 2002).

10.4. Findings

In the case of Equinor, a sense of moral legitimacy enables owners, the board of directors and a key manager to reshape the company's strategy in order to take a step towards a new, sustainable transition (Figure 10.1). Pragmatic legitimacy enables the company's employees to have a self-interest in the company's sustainable future and to facilitate future sustainable innovation. Finally, cognitive legitimacy requires internal stakeholders to understand and accept the purpose of the company's new sustainable strategy so it will be accepted by the public.

First, moral legitimacy identifies the reformulation of sustainability strategy that integrates new innovative transitions at Equinor. Second, pragmatic legitimacy identifies the creation of the new innovative solutions (such as offshore wind energy projects). Third, cognitive legitimacy identifies the commercial interest of the offshore wind project that was promoted by the sustainability strategy. Thus, three types of legitimacy enabled Equinor to develop new clean alternative sources of energy and, more specifically, to enter a new offshore wind market.

10.4.1 Dimensions of legitimacy

The three types of legitimacy used in this study were defined in accord with Idowu *et al.* (2013) and Suchman (1995), and expressed as follows.

Figure 10.1 The role of legitimacy in creating SD transition at Equinor

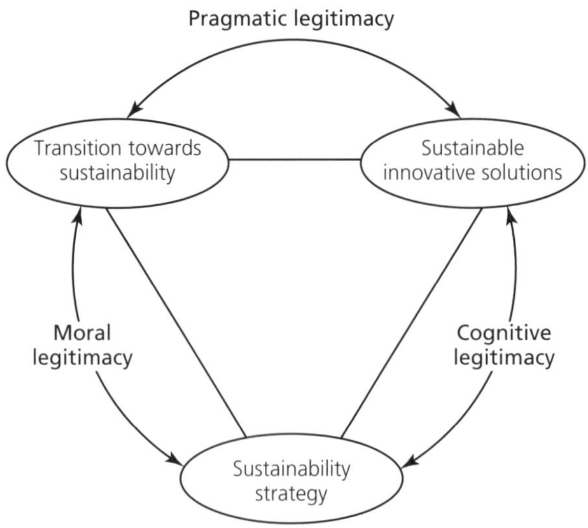

Table 10.1 Prevalence of code hierarchies

Type of legitimacy	Sources	References
Moral legitimacy	43	252
Pragmatic legitimacy	39	199
Cognitive legitimacy	21	45

Moral legitimacy: 'the right thing to do'.

- Norway has a long-term SD future (e.g. following the Bruntland Report).
- Equinor's owners (70% state owned), board of directors and 'top leader' reshaped the company's strategy to achieve a SD future.
- A clear SD transition is introduced.

Pragmatic legitimacy: self-interested direct exchange between the company and its internal stakeholders; political, economic, and social interdependencies that affect employees' views about SD transition.

- RE activities are perceived as a business case, and have a specific interest for the company's internal stakeholders.
- People in the company are involved in the new transition process.
- People in the company are interested to use their competence and knowledge in the new transition process.

Cognitive legitimacy: comprehensible, permanence, plausibility – which is adopted and taken for granted.

- A strategic driving force is created to achieve new value propositions (economic, environmental and social).
- A clear competitive strategy is developed.
- RE is mostly taken for granted by internal stakeholders.

Table 10.1 shows the distribution of the three examined forms of legitimacy.

10.4.2 Moral legitimacy to SD transition: embracing the 'right thing to do'

Table 10.1 indicates the prevalence of moral legitimacy found in the data. But this quantitative measure does not adequately demonstrate the complexity of moral legitimacy expression by the respondents. The following quotations provide a flavour of the interview data, focusing on moral legitimacy.

Interviewees suggested that the move towards SD at Equinor came from new management. For instance

> What has changed is maybe the change of the CEO and change of market, and right now these initiatives have a higher profile than they had 2 years ago. So the question to ask is ... what shapes Statoil's image? Is it the project pipeline or is it the strategy on this asset versus that asset. So that may be a better framing, and you will see that the new narrative towards greener energy and lower carbon has really taken off with [new manager].
>
> (Head of Corporate Strategy Development, Equinor, 2015)

Moreover

> As stated by our top leader Eldar Sætre, we need to take a responsibility as an OG company in order to acknowledge that we are part of the climate issue.
>
> (Senior Sustainability Consultant, Equinor)

As the government is the major owner of the company, its opinion weighs heavily when selecting a new CEO; senior management is thus aligned with government.

> I think a lot of senior management is born and bred in oil and gas, and it will be hard to see Statoil become something else. But then I see a strategic mind-set and people who are more oriented towards other businesses – they see that this is coming, in some form at some point. So I think this is not a democratic decision, this is the board and the owner, and [new manager] and probably a few that are trusted. He made the decision to establish new energy solutions, and making that executive, and they reported to him before going lower down in the organisation.
>
> (Head of Innovation Special Projects, Equinor, 2015)

Consequently, the move towards SD was mainly seen as an extension of the government's policy towards a low-carbon future.

> On the government's side, the government has a couple of people who sit on the board of directors but they don't actively manage or steer the strategy of the company – Norwegian capitalism works very well. They are not implying what we should do, or telling us what we need to do, or anything like this: you could continue to do oil and gas – if you want that, it is OK. But there is a feeling that they want Statoil to take leadership on this issue, particularly on transition – slow transition from oil and gas towards other areas.
>
> (Senior Vice President, Equinor, 2016)

The acceptance of this new strategy by the interviewees reflects the belief that the company should adhere to pressure from government and public opinion.

> I think there are multiple overlapping agendas here. Political pressure – I think there is a genuine acknowledgement that we are part of the problem here, that we are producing a lot of carbon emissions, or that our product is a carbon product,

> which hits the climate hard. I think it is also seeing new energy sources coming in to compete with oil and gas. And everything is connected with the political agenda. Our owner is the state, basically the biggest owner is the Norwegian government. So if they decide that they want to pursue this agenda, do we really have a choice? But sometimes people make it too simple and say it is one reason why you do something, but I think it is multiple reasons, and probably some of them are better than others.
>
> (Head of Innovation Special Projects, Equinor, 2015)

We also see in the interviews a clear sign that the respondents are happy that Equinor makes an effort to 'do the right thing' with regards SD strategy and investment in renewables.

> Two aspects to that: one, this pressure we have in Norway is making us try to be the most efficient possible, which pushes us to improve our performance internally; two, we have the portfolio that we have today, to have a lower impact on the environment and climate side. We might need to do things differently, and include more renewables in our portfolio.
>
> The reason why Statoil went into renewables is due to, I think, that it didn't want to be a part of the problem but a part of the solution.
>
> (Strategy Innovation Manager, Equinor, 2015)

People at Equinor support RE, but they also believe that the energy market is changing.

> And what we see at the moment with Statoil's interest in renewables is driven not only by government owners but by the willingness and the change that are happening in the oil and gas business.
>
> (Responsible for Wind Operations, Equinor, 2016)

Moreover, a low carbon dioxide footprint will have the potential to open new business opportunities for Equinor. This relates to another level of moral legitimacy, as discussed above (see Section 10.2.2), to do with ensuring the future of the company, and the goods associated with what a large corporation can do for employees, the wider population and the nation as a whole.

> So I think for a carbon-producing country Norway has an interest to ensure that Statoil succeeds in the long term. To produce good energy for the country, jobs and the financial benefits.
>
> (Head of Corporate Strategy Development, Equinor, 2015)

Shifting Equinor from an OG company to a broader mixed-energy company was supported by a normative assessment by the owners, the board of directors and the new CEO of Equinor. Statements such as those above indicate that moral legitimacy played an essential role in triggering the introduction of RE in Equinor – and is closely linked to a longer-term sustainability strategy. These findings are supported by the studies of Ajzen (1985, 1991) and Thomas and Lamm (2012).

10.4.3 Pragmatic legitimacy to SD transition: self-interest for the company

The previous section showed a moral acceptance of the SD transition at Equinor, but there is less clarity about what it means in practical terms, with employees having differing opinions about the routes to a low-carbon future.

> It depends who you talk to ... With some exceptions you see a very big interest, especially at the lower level ... Some want to contribute and some are afraid to jump into something new.
>
> (Strategy and Innovation Manager, Equinor, 2015)

Thus, some employees supported the wider sustainable shift in Equinor, while others believed that the company should focus only on 'greening' OG production since it brings more profit to the company. Still others suggested that the sustainable strategic move would make it easier to pursue the company's traditional business; something that gives credence to greenwashing motives.

> Statoil has historically for several decades kept an eye on the outskirts of its activities, because that's where innovation and technology development often takes place. So my suspicion is that the investment in renewables is a part of Statoil's strategy to build a reputation which makes it easier to access oil and gas acreage in Northern Norway.
>
> (Former Chief Economist, Equinor, 2015)

The strategic move was seen as an investment tactic, not necessarily implemented for a belief in SD per se.

> It could be because they believe in it, or it can be because they see we are moving from oil and gas and this is the new place to be, I don't know. I think you also have another opinion in the organisation that is 'Yes, these renewables – we have tried them before, maybe this will blow over and we will go back to what we've always done and can do'.
>
> (Head of Innovation Special Projects, Equinor, 2015)

However, others believe in SD transition as a sincere route to another important business opportunity.

> So it is good to have a mixture, so you become more robust, and can play a different role in different sources.
>
> (Strategy and Innovation Manager, Equinor, 2015)

We found that the interviewees displayed diverse views of how the new way forward could be of interest to the company itself. We observed their ambiguity as a translation process, where they grappled with the meaning of the new sustainability strategy. As a result, we claim that pragmatic legitimacy helps employees to approach SD transition. These findings are supported by prior studies of legitimacy and organisations (Idowu *et al.*, 2013). In addition, according to Meijer and Bolivar (2016), smart governance involves a complex process of institutional change, but where a majority of employees come to understand the current debate about the SD transition, and believe in this transition.

10.4.4 Cognitive legitimacy: do internal stakeholders have the knowledge to be in this business and what is the public perception of the company's actions?

Equinor aims to deliver profitable projects to survive, remain attractive and gain wider public acceptance. Not surprisingly, Equinor's chief economist notes that

> The key driver of any investment decision is economic value and net present value. That's how it used to be and how it will be in the future.
>
> (Former Chief Economist, Equinor, 2015)

It therefore also follows that

> Statoil will have no social acceptance if we are doing RE projects where we are losing money.
>
> (Special Consultant for Exploration Activities, Equinor, 2017)

But this is within context in which

> There will be more and more need for energy, whether from OG or cleaner energy from RE ... We know that OG are not here forever, they are finite resources ... We need to invest or use these revenues to create new opportunities for the next generation.
>
> (Senior Sustainability Consultant, Equinor, 2014)

However, while Equinor is seeking to provide sustainable energy for the future by investing in a new low-carbon business (RE), this activity is outside its core business, and this raises cognitive legitimacy issues.

> I think RE is producing a lot of questions at the moment, because it is different ... OG is an extraction industry, renewables are a production, manufacturing industry ... So we are positioning ourselves in a different industry sector.
>
> (Project Development Department, Equinor)

On the other hand, Equinor leaders recognise that the company has transferable experience, skills and knowledge from OG offshore activities that lend themselves to SD, such as offshore wind.

> They see that we can utilise our skills that have been built up over the last 40 years, in broadening our activities into RE ..., and we have a lot of competence in everything we do offshore.
>
> (Head of Wind Assets Management, Equinor)

> We use the skills that we have from OG. They are very transferable skills, especially offshore wind, marine operations, large projects, putting massive capital expenditure projects together – it's the stuff we do in OG.
>
> (Senior Vice President, Equinor, 2016)

In addition, according to Equinor's public website, the company believes that the passion and curiosity of its employees will guide them in how to do business, and that will enable the transition to a new energy future (Statoil, 2017). Equinor has invested in internal dialogues to develop thinking around SD by

> encouraging people to come forward with new ideas and [through] various initiatives for people to suggest solutions.
>
> (Project Development Department, Equinor)

At the same time, for employees to perceive offshore wind projects as, for example, business cases, such projects must include positive economic, environmental and social impacts for local communities.

> It's the energy to people and the clean energy to people.
>
> (Vice President, Equinor, 2017)

This is essential for the internal stakeholders to understand in order to accept the new shift of the company.

Equinor claims that it works to find clean alternative sources of energy such as RE as part of its environmental and social obligation.

> I think it has more to do with sustainability and creating an environment that is sustainable for generations to come.
>
> (Responsible for Wind Operation, Equinor, 2016)

Therefore, Equinor intended to enter the RE market not only because it is good for the company and its stakeholders but also good for the environment and good for societies.

We claim that cognitive legitimacy involves its employees in the new SD transition by investing in potentially profitable projects. Equinor also believes its employees will guide them in the transition to a new energy future (Statoil, 2017). In addition, for employees to use their skills to develop new offshore wind innovations, projects must include social, environmental and economic factors. These factors support a sustainability cognitive legitimacy that is comprehensible and taken for granted by the company's internal stakeholders – an automatic adaptation to social expectations (Idowu *et al.*, 2013; Suchman, 1995).

10.5. Discussion and implications

The power industry faces fundamental changes in the European market due to change within transportation (e.g. electric vehicles), in addition to the increasing demand for electricity from renewables (Adhi Setiawan and Asvial, 2016; Cheng *et al.*, 2017). Furthermore, environmental regulations play important roles in forcing companies to move towards SD actions (Hoffman, 1999; Richards and Gladwin, 1999). Many companies have considered sustainability as an important tool to integrate economic, environmental and social dimensions, and are increasingly addressing sustainability as a core part of every decision they make.

Equinor is an interesting example of a company whose profitability aims and whose primary operation would appear to be at odds with sustainability developments. However, various contexts – including the Norwegian government's commitment to SD and Equinor's economic interest in diversification to renewables – have influenced change in the company. Equinor's first pilot of an offshore floating wind turbine was established in Norway in 2008 (Equinor, 2008). However, according to one of our interviewees, the primary aim of this project may not have been to start a new clean energy solution but to enhance Equinor's reputation, and to establish better market access in Northern Norway. In our opinion, these are perhaps best seen as competing priorities: in the context of the Norwegian government's commitment to RE, Equinor needed to expand its SD investment.

In general, an OG company cannot readily adapt new sustainable business in a short time frame. Indeed, fundamental changes must occur in the company's longer-term strategy in order to achieve a sustainable business future. In order to better understand how SD comes to be implemented within the company's strategy for the future, we have explored the role of legitimacy within Equinor. Each type of legitimacy plays a different role in implementing SD.

- **Moral legitimacy.** According to (Deegan, 2002), moral legitimacy is essential: managers engage in legitimisation strategies to ensure the survival of the company (Deegan, 2002). In the case of part-government-owned Equinor, the introduction of a new sustainable business is facilitated by the government's strict climate policy supporting SD transition, and through the board of directors and the employment of a CEO who decided to invest heavily in RE, thereby leading Equinor's 'green shift' from purely OG to broader energy.
- **Pragmatic legitimacy.** Palazzo and Scherer (2006) argue that pragmatic legitimacy is engaged in by different stakeholders to help ensure access to resources and

markets, and lead to the continuity of the company. In the case of Equinor, many employees perceived a new sustainable transition as a business opportunity; however, the translation process of SD will continue within the company until the SD transition is complete.

- **Cognitive legitimacy.** Idowu *et al.* (2013) argued that cognitive legitimacy is the visible symbol of legitimacy. In the case of Equinor, employees are engaged in the SD transition process. This assists employees' acceptance of a shift to renewables, because they understand the changing market and believe in their competence to provide citizens with RE products that have social, environmental and economic benefits.

Cognitive legitimacy is difficult to assess since it aims to consider acceptance – perceived legitimacy – by outside parties. The first stage of this is employees' recognition that they have the knowledge and skills to do the job; the second – though not the subject of this chapter – is how company policy and actions are perceived by the wider public (Idowu *et al.*, 2013), which has been termed a 'social licence to operate' and appears to be a new name for legitimacy that means 'legitimating a company's action' (Gehman *et al.*, 2017). The latter is particularly difficult to ascertain for methodological reasons but also, in this instance, because of the role of the media, and public opinion as reported in the mainstream media, regarding the importance of SD.

Our evidence suggests that in the case of Equinor, moral legitimacy contributed most to the introduction of sustainable activities. This was related to the direct support from the Norwegian government and the board of directors, and the new CEO's concern to encourage a broader energy mix. However, our focus on pragmatic legitimacy showed us different views of the employees, where doubts about SD surfaced alongside different expressions of self-interest in translating a sustainability agenda into action. Finally, our consideration of cognitive legitimacy showed us that a majority of employees accepted the SD transition, at least in part due to recognising their ability to use their existing offshore skills in order to develop offshore wind projects, and due to coming to understand the relevance of the SD transition in context and to accept its social and environmental benefits. However, in terms of the economic benefit, Equinor should expand its business in RE in order to be accepted by all parties in the company. This makes cognitive legitimacy challenging to measure at this moment in time.

10.5.1 Implications for government, decision-makers and the OG industry

Governance plays the central role in bringing smart city initiatives to citizens (Belissent, 2010).

In addition, government in smart governance perspective plays a critical role in fostering smart cities (Yigitcanlar *et al.*, 2008). For example, Meijer and Bolívar (2013) believed that solving societal problems is not about developing good policies but it is about a managerial role that organises strong collaborations between government and other stakeholders in order to better serve citizens and communities.

Norwegian OG production remains central to Norway's economy (Thurber and Istad, 2011). However, as part of the Norwegian government's long-term SD agenda,

Norway introduced carbon dioxide taxes as a step to limit carbon emissions (KLD, 2014). In addition, Norway is working towards the Paris Agreement to ensure a reduction in carbon emissions (Thurber and Istad, 2011; UN, 2017). This led the Norwegian government to include sustainability into the core of Equinor's business.

At the same time, Equinor assumes that the demand for energy, including OG, will increase in the foreseeable future (Bjartnes, 2015), which makes it an obvious option to continue extracting OG. Equinor considers its production of OG to be efficient and of a high environmental standard for the sector, and believes that if it stops producing OG, competitors will produce it and cause more damage to the environment.

However, in order to be part of the solution rather than part of the problem, Equinor has invested in offshore wind power, increasingly outside Norway. This is due to the fact that Norway's renewables market is based on hydropower (KLD, 2014), making it difficult for Equinor to compete. While Equinor has mainly established its RE projects outside Norway, this may still provide significant benefits for Norway. For example, Norwegian companies are responsible for implementing the engineering work needed to build the wind projects. In addition, the profit gained from RE projects contributes to further increased economic growth in Norway.

10.5.2 Limitations and implications for further research

Although it is difficult to draw generalisations from only one case, the richness of the semi-structured interviews helped us better understand the role of legitimacy in introducing RE activities in an OG company. Further research, using larger samples of OG companies and other sectors, is needed to confirm the results of this study and generalise its conclusions. In addition, to uncover the role of legitimacy in introducing sustainable activities, it is necessary to understand the dynamic process of legitimacy in a broader context; that is, in other sectors and regions.

10.6. Conclusion

The demand for energy, and particularly clean energy, is predicted to increase in the coming decades. This chapter presented a case study of a state-owned OG company that intended to be part of the solution and transitioned to broad mixed-energy production by investing heavily in clean energy activities, including, notably, off-shore wind power. This chapter explored how this transition occurred in Equinor, using the concept of legitimacy to help us gain insight into the complex process of a SD transition.

'Smart governance' requires engagement with stakeholders to collaborate to establish companies' legitimacy and sustainability. The implications of this are considerable: where energy sources are more sustainable, and where there is robust support for SD from government, companies, employees and citizens, there is greater opportunity to create smarter cities (and greener futures).

Acknowledgements

We would like to thank Dr David Gibson and Associate Professor Matthew W. Rutherford for their valuable feedback on this chapter. We would also like to thank Mrs

Natacha Blisson and Mr Morten Mikkelsen from Equinor for their assistance, advice and answering our many questions, as well as the interviewees for sharing their stories and insights.

REFERENCES

Adhi Setiawan E and Asvial M (2016) Renewable energy's role in a changing world. *International Journal of Technology* **7(8)**: 1280–1282.

Ajzen I (1985) From intentions to actions: a theory of planned behavior. In: *Action Control* (Kuhl J and Beckmann J (eds)). Springer, Berlin, Germany.

Ajzen I (1991) Theories of cognitive self-regulation. The theory of planned behavior. *Organizational Behavior and Human Decision Processes* **50**: 179–211.

Albino V, Berardi U and Dangelico RM (2015) Smart cities: definitions, dimensions, performance, and initiatives. *Journal of Urban Technology* **22**: 3–21.

Aldrich HE and Fiol CM (1994) Fools rush in? The institutional context of industry creation. *The Academy of Management Review* **19**: 645–670.

Almutairi A, Gardner G and McCarthy A (2013) Practical guidance for the use of a pattern-matching technique in case-study research: a case presentation. *Nursing and Health Sciences* **16**: 239–244.

Azmat F and Samaratunge R (2009) Responsible entrepreneurship in developing countries: understanding the realities and complexities. *Journal of Business Ethics* **90**: 437–452.

Batagan L (2011) Smart cities and sustainability models. *Informatica Economica* **15**: 80–87.

Belissent J (2010) *Getting Clever About Smart Cities: New Opportunities Require New Business Models*. Forrester Research, Cambridge, MA, USA.

Beveridge WIB (1951) *The Art of Scientific Investigation*. William Heinemann, London, UK.

Binz C, Harris-Lovett S, Kiparsky M, Sedlak DL and Truffer B (2016) The thorny road to technology legitimation – institutional work for potable water reuse in California. *Technological Forecasting and Social Change* **103**: 249–263.

Bitektine A (2011) Toward a theory of social judgments of organizations: the case of legitimacy, reputation, and status. *Academy of Management Review* **36**: 151–179.

Bjartnes A (2015) Statoil's big dilemma: should it continue to go for oil and gas – or transform itself into an energy service provider? Energy Post. http://energypost.eu/statoils-big-dilemma-continue-go-oil-gas-transform-energy-service-provider (accessed 20/06/2019).

Bloodgood J, Hornsby J, Rutherford M and McFarland R (2017) The role of network density and betweenness centrality in diffusing new venture legitimacy: an epidemiological approach. *International Entrepreneurship and Management Journal* **13**: 525–552.

Bridwell-Mitchell EN and Mezias SJ (2012) The quest for cognitive legitimacy: organizational identity crafting and internal stakeholder support. *Journal of Change Management* **12**: 189–207.

Brundtland GH (1987) *Report of the World Commission on Environment and Development: Our Common Future*. Oxford University Press, Oxford, UK.

Calvillo CF, Sánchez A and Villar J (2013) Distributed energy generation in smart cities. *International Conference on Renewable Energy Research and Applications (ICRERA), Madrid, Spain*, pp. 161–166.

Cheng X, Korpås M and Farahmand H (2017) The impact of electrification on power system in Northern Europe. *14th International Conference on the European Energy Market (EEM 2017), Dresden, Germany*, pp. 1–7.

Clark SM, Gioia DA, Ketchen DJ and Thomas JB (2010) Transitional identity as a facilitator of organizational identity change during a merger. *Administrative Science Quarterly* **55**: 397–438.

Creswell JW (2012) *Qualitative Inquiry and Research Design: Choosing Among Five Approaches*, 3rd edn. Sage, London, UK.

Csomós G (2014) Relationship between large oil companies and the renewable energy sector. *Environmental Engineering and Management Journal* **13**: 2781–2787.

Czinkota M, Kaufmann HR and Basile G (2014) The relationship between legitimacy, reputation, sustainability and branding for companies and their supply chains. *Industrial Marketing Management* **43**: 91–101.

Deegan C (2014) An overview of legitimacy theory as applied within the social and environmental accounting literature. In: *Sustainability accounting and accountability* (Bebbington J, Unerman J and O'Dwyer B (eds)), 2nd edn. Routledge, London, UK.

Deegan C (2002) Introduction: The legitimising effect of social and environmental disclosures – a theoretical foundation. *Accounting, Auditing and Accountability Journal* **15**: 282–311.

Diesendorf M (2000) Sustainability and sustainable development. In: *Sustainability: The Corporate Challenge of the 21st Century* (Dunphy D, Benveniste J, Griffiths A and Sutton P (eds)). Allen and Unwin, Sydney, Australia.

Dowling J and Pfeffer J (1975) Organizational legitimacy: social values and organizational behavior. *The Pacific Sociological Review* **18**: 122–136.

Du S, Bhattacharya CB and Sen S (2007) Reaping relational rewards from corporate social responsibility: the role of competitive positioning. *International Journal of Research in Marketing* **24**: 224–241.

Dyer W, Wilkins A and Eisenhardt K (1991) Better stories, not better constructs, to generate better theory: a rejoinder to Eisenhardt. *Academy of Management Review* **16(3)**: 613–619.

Eisenhardt KM and Graebner ME (2007) Theory building from cases: opportunities and challenges. *The Academy of Management Journal* **50**: 25–32.

Elgazzar RF and El-Gazzar RF (2017) Smart cities, sustainable cities, or both? A critical review and synthesis of success and failure factors. *The 6th International Conference on Smart Cities and Green ICT Systems (SMARTGREENS 2017), Poerto, Portugal*, pp. 250–257.

Engert S, Rauter R and Baumgartner RJ (2016) Exploring the integration of corporate sustainability into strategic management: a literature review. *Journal of Cleaner Production* **112**: 2833–2850.

England P (2009) Managing urban water in Australia: the planned and the unplanned. *Management of Environmental Quality: An International Journal* **20**: 592–608.

Epstein MJ and Roy MJ (2001) Sustainability in action: identifying and measuring the key performance drivers. *Long Range Planning* **34**: 585–604.

Equinor (2019a) Wind farms were considered a threat to national security. https://www.equinor.com/en/magazine/offshore-wind-radar-mitigation.html (accessed 20/06/2019).

Equinor (2019b) Capturing renewable business opportunities. https://www.equinor.com/en/what-we-do/new-energy-solutions.html (accessed 20/06/2019).

Equinor (2008) StatoilHydro to build first full scale offshore floating wind turbine. https://www.equinor.com/en/news/archive/2008/05/22/hywindfullscale.html (accessed 20/06/2019).

European Commission (2016) Endorse the Basque Declaration. http://www.sustainablecities.eu/endorse-the-basque-declaration (accessed 20/06/2019).

Fisher J and Bonn I (2011) Sustainability: the missing ingredient in strategy. *Journal of Business Strategy* **32**: 5–14.

Flyvbjerg B (2006) Five misunderstandings about case-study research. *Qualitative Inquiry* **12**: 219–245.

Frandsen S, Morsing M and Vallentin S (2013) Adopting sustainability in the organization: managing processes of productive loose coupling towards internal legitimacy. *Journal of Management Development* **32**: 236–246.

Freudenburg WR and Gramling R (2011) *Blowout in the Gulf: The BP Oil Spill Disaster and the Future of Energy in America*. MIT Press, Cambridge, MA, USA.

Gehman J, Lefsrud LM and Fast S (2017) Social license to operate: legitimacy by another name? *Canadian Public Administration* **60**: 447–447.

Gibbs G (2002) *Qualitative Data Analysis: Explorations with NVivo*. Open University Press, Buckingham, UK.

Giffinger R, Fertner C, Kramar H *et al.* (2007) *Smart Cities: Ranking of European Medium-sized Cities*. Vienna University of Technology, Vienna, Austria. http://www.smart-cities.eu/download/smart_cities_final_report.pdf (accessed 20/06/2019).

Gioia Da, Schultz M and Corley KG (2000) Organizational identity, image, and adaptive instability. *The Academy of Management Review* **25**: 63–81.

Guerreiro A (2015) Impact of IS/IT investments on firm performance: does stakeholder orientation matter? *Electronic Journal Information Systems Evaluation* **19(2)**: 99–111.

Handelman JM and Arnold SJ (1999) The role of marketing actions with a social dimension: appeals to the institutional environment. *Journal of Marketing* **63**: 33–48.

Hoffman AJ (1999) Institutional evolution and change: environmentalism and the U.S. chemical industry. *The Academy of Management Journal* **42**: 351–371.

Hsieh HF and Shannon SE (2005) Three approaches to qualitative content analysis. *Qualitative Health Research* **15(9)**: 1277–1288.

Hughes C and Spray R (2002) Smart communities and smart growth – Maximising benefits for the corporation. *Journal of Corporate Real Estate* **4**: 207–214.

Idowu SO, Capaldi N, Zu L and Gupta AD (2013) *Encyclopedia of Corporate Social Responsibility*. Springer, Berlin, Germany.

Jacobsen DI, Sandin G and Hellström C (2002) *Vad, hur och varför: om metodval i företagsekonomi och andra samhällsvetenskapliga ämnen*. Studentlitteratur, Lund, Sweden.

Jepperson RL (1991) Institutions, institutional effects, and institutionalism. In: *The New Institutionalism in Organizational Analysis* (Powell WW and DiMaggio PJ (eds)). University of Chicago Press, Chicago, IL, USA.

Kishna M, Niesten E, Negro S and Hekkert MP (2017) The role of alliances in creating legitimacy of sustainable technologies: a study on the field of bio-plastics. *Journal of Cleaner Production* **155**: 7–16.

Kitchin R (2015) Making sense of smart cities: addressing present shortcomings. *Cambridge Journal of Regions, Economy and Society* **8**: 131–136.

KLD (Norwegian Ministry of Climate and Environment) (2014) *Norway's Sixth National Communication, Under the Framework Convention on Climate Change*. KLD, Oslo, Norway.

Kotler P and Lee NR (2005) *Corporate Social Responsibility: Doing The Most Good for Your Company and Your Cause*. Wiley, Hoboken, NJ, USA.

Kumar V, Rahman Z, Kazmi AA and Goyal P (2012) Evolution of sustainability as marketing strategy: beginning of new era. *Procedia – Social and Behavioral Sciences* **37**: 482–489.

Lamin A and Zaheer S (2012) Wall Street vs. main street: firm strategies for defending legitimacy and their impact on different stakeholders. *Organization Science* **23**: 47–66.

Lee JH, Phaal R and Lee SH (2013) An integrated service–device–technology roadmap for smart city development. *Technological Forecasting and Social Change* **80**: 286–306.

Ludlow D (2016) *Shaping Common Ground in Urban Sustainability? Smart City Governance Common Ground for Integrated Urban Management*. Urban Europe, Wien, Austria. http://jpi-urbaneurope.eu/app/uploads/2017/04/Ludlow_Shaping_common_ground_in_urban_sustainability__Smart_city_governance.pdf (accessed 20/06/2019).

Meijer A and Bolívar MPR (2013) Governing the smart city: scaling-up the search for socio-techno synergy. *The European Group For Public Administration (EGPA) Conference, Edinburgh, UK*.

Meijer A and Bolívar MPR (2016) Governing the smart city: a review of the literature on smart urban governance. *International Review of Administrative Sciences* **82**: 392–408.

Miles MB, Huberman MA and Saldana J (1994) *Qualitative Data Analysis*. Sage, Thousand Oaks, CA, USA.

Ødegaard ARS and Grønning M (2017) *Smart, Social and Sustainable?: Norwegian Smart Cities; Means for Social Sustainability or Expressions of a Techno-economic Race?* Masters thesis, Norwegian University of Life Sciences, Ås, Norway.

Palazzo G and Scherer AG (2006) Corporate legitimacy as deliberation: a communicative framework. *Journal of Business Ethics* **66**: 71–88.

Parsons T (1960) Structure and process in modern societies. *American Journal of Sociology* **66**: 5.

Patton MQ (2002) *Qualitative Research and Evaluation Methods*. Sage, Thousand Oaks, CA, USA.

Pellicer S, Santa G, Bleda AL *et al.* (2013) A global perspective of smart cities: a survey. *Seventh International Conference on Innovative Mobile and Internet Services in Ubiquitous Computing, Taichung, Taiwan*, pp. 439–444.

Petkova AP (2016) Standing out or blending in? The formation of new firms' legitimacy and reputation under different levels of market uncertainty. *Corporate Reputation Review* **19**: 22–34.

Richards DJ and Gladwin TN (1999) Sustainability metrics for the business enterprise. *Environmental Quality Management* **8**: 11–21.

Sarker TK (2013) Voluntary codes of conduct and their implementation in the Australian mining and petroleum industries: is there a business case for CSR? *Asian Journal of Business Ethics* **2**: 205–224.

Scott WR (2014) *Institutions and Organizations: Ideas, Interests, and Identities*. Sage, Thousand Oaks, CA, USA.

Statoil (2017) Careers. https://www.statoil.com/en/careers.html (accessed 20/06/2019).

Suchman MC (1995) Managing legitimacy: strategic and institutional approaches. *Academy of Management Review* **20(3)**: 571–610.

Taherdangkoo M, Ghasemi K and Beikpour M (2017) The role of sustainability environment in export marketing strategy and performance: a literature review. *Environment, Development and Sustainability* **19**: 1601–1629.

Thomas T and Lamm E (2012) Legitimacy and organizational sustainability. *Journal of Business Ethics* **110**: 191–203.

Thurber MC and Istad BT (2011) Norway's evolving champion: Statoil and the politics of state enterprise. In: *Oil and Governance: State-Owned Enterprises and the World Energy Supply* (Victor DG, Hults DR and Thurber MC (eds)). Cambridge University Press, Cambridge, UK.

Timothy G, Lee WJ and Greg B (2015) Is your sustainability strategy sustainable? Creating a culture of sustainability. *Corporate Governance: The International Journal of Business in Society* **15**: 1–17.

Tolbert P and Zucker L (1983) Institutional sources of change in the formal structure of organizations: the diffusion of Civil Service reform, 1880–1935. *Administrative Science Quarterly* **28**: 22.

UN (United Nations) (2017) *The Paris Agreement*. UN, Geneva, Switzerland. http://unfccc.int/paris_agreement/items/9485.php (accessed 20/06/2019).

UN (1987) *Our Common Future: Report of the World Commission on Environment and Development*. UN, Geneva, Switzerland. https://sustainabledevelopment.un.org/content/documents/5987our-common-future.pdf (accessed 20/06/2019).

Vallance S, Perkins HC, Bowring J and Dixon JE (2012) Almost invisible: glimpsing the city and its residents in the urban sustainability discourse. *Urban Studies* **49**:1695–1710.

Van de Ven AH and Poole MS (2002) *Field Research Methods*. In: *Companion to Organizations* (Baum JAC (ed.)). Blackwell, Oxford, UK, pp. 867–888.

Vanolo A (2014) Smartmentality: the smart city as disciplinary strategy. *Urban Studies* **51**: 883–898.

Wan Ahmad WNK, Rezaei J, De Brito MP and Tavasszy LA (2016) The influence of external factors on supply chain sustainability goals of the oil and gas industry. *Resources Policy* **49**: 302–314.

Woetzel J and Jaana R (2018) Is your company ready for the rise of smart cities? *Harvard Business Review*, 12 June.

Yigitcanlar T, Kamruzzaman M, Buys L *et al.* (2018) Understanding 'smart cities': intertwining development drivers with desired outcomes in a multidimensional framework. *Cities* **81**: 145–160.

Yigitcanlar T, Velibeyoglu K and Martinez-Fernandez C (2008) Rising knowledge cities: the role of urban knowledge precincts. *Journal of Knowledge Management* **12**: 8–20.

Yin RK (2014a) Collecting case study evidence: the principle you should follow in working with six sources of evidence. In: *Case Study Research: Design and Methods*, 5th edn. Sage, Los Angeles, CA, USA.

Yin RK (2014b) Getting started: how to know whether and when to use the case study as a research method. *Case Study Research: Design and Methods*, 5th edn. Sage, Los Angeles, CA, USA.

Yin RK (1994) *Case Study Research: Design and Methods*. Sage, Thousand Oaks, CA, USA.

Yoon Y, Gürhan-Canli Z and Schwarz N (2006) The effect of corporate social responsibility (CSR) activities on companies with bad reputations. *Journal of Consumer Psychology* **16**: 377–390.

Zhang Y and Wildemuth BM (2009) In applications of social research methods to questions in information and library science. In: *Qualitative Analysis of Content* (Wildemuth B (ed.)). Libraries Unlimited, Cullompton, UK.

Zimmerman MA and Zeitz GJ (2002) Beyond survival: achieving new venture growth by building legitimacy. *The Academy of Management Review* **27**: 414–431.

ice Publishing

Nuttall, William J, Gibson, David G, Trzmielak, Dariusz and Ibarra-Yúnez, Alejandro
ISBN 978-0-7277-6425-6
https://doi.org/10.1680/emsc.64256.211

Chapter 11
The crucial role of citizen involvement in smart city development and operation

Richard Bull
Nottingham Trent University, UK

Ken Dooley
Aalto University, Finland

Muhammad Mazhar
Nottingham Trent University, UK

> Sicinius: What is the city but the people?
>
> Citizens: True, the people are the city.
>
> William Shakespeare, *Coriolanus*, Act 3, Scene 1

11.1. Introduction

A significant cultural shift occurred recently with the majority of the world's population now living in cities and contributing over two-thirds of global carbon emissions (UNEP, 2015). If countries such as the UK are to meet their challenging carbon reduction targets (80% by 2050 for the UK), then how our cities are governed and managed to maximise energy efficiency is of vital importance. Faith is increasingly being placed in what are commonly referred to as 'smart cities' to meet these targets. Most visions of these smart cities revolve around increased information and communications technology (ICT) efficiency through what has become known as the 'digital economy'. Smart cities seemingly offer a utopian vision of urban integration, efficiency and subsequent carbon reductions, yet urbanisation presents real challenges, as noted by the fact that smart cities now features as a United Nations (UN) Sustainable Development Goal. Smart cities and communities is Sustainable Development Goal 11: 'Make cities and human settlements inclusive, safe, resilient and sustainable.' Carbon reductions and environmental considerations are just one challenge for future cities. These densely populated urban centres pose significant resource challenges for energy, water and food, and for transport, planning and infrastructure. The UN (2018) notes that

> Rapid urbanization brings enormous challenges, including growing numbers of slum dwellers, increased air pollution, inadequate basic services and

infrastructure, and unplanned urban sprawl – which also make cities more vulnerable to disasters.

In response to these challenges, both technology giants such as Schneider, Cisco and Siemens and policy-makers believe that the opportunities afforded by integrated data platforms to connect energy, water and transport will transform our cities. But is 'smart' purely seeking maximum technical efficiencies or does 'smart' need to incorporate citizens as well? Cities, we argue (borrowing a well-cited phrase from Janda (2011)), like any building development or infrastructure, do not use energy – people do. Concerns have been raised by academics (Cowley and Caprotti, 2018; Martin *et al.*, 2018) that such interpretations of smart cities are lacking a democratic mandate and also perpetuate a consumerist growth agenda that will fail to resolve the underlying problems facing cities.

Whether the smart city is real or just a marketing opportunity by the global tech giants, the market is real and growing at a huge pace. Future Cities Catapult (2017) noted that the projected market context for smart cities is set to more than double over the next few years, from approximately $300 billion in 2015 to $750 billion in 2020, with particularly high growth expected in Asia Pacific.

This chapter first critically explores definitions of smart cities before considering the academic literature surrounding smart cities and citizen engagement. Secondly, two aspects of smart cities and citizen engagement are discussed – one is a short case study of an EU lighthouse project in Nottingham, UK, that is contrasted with novel forms of digital engagement with a particular consideration for how connected citizenship is evolving.

11.2. Smart cities – an evolving concept

The phrase 'smart city' has emerged over the last 25 years, and has been used predominantly by the ICT sector and companies such as IBM, Cisco and Siemens. Definitions of smart cities vary according to the sector in which they are used, and it is immediately evident from the range of definitions that there is little consensus. The range of industrial definitions were chronicled by Bull and Azzenoud (2016) and are listed in Table 11.1.

Table 11.1 reflects the first stage of what Future Cities Catapult referred to as the 'marketeer's' vision of smart cities, which they felt dominated in the 1990s. The focus was on capitalising on the potential of ICT solutions to connect energy, water and transport. At this stage the term 'smart city' would have been interchangeable with the 'information' or 'digital' city. From here, commentators note a second stage with visions and definitions expanding to include citizen engagement in various forms – be it face-to-face participatory processes or on-line engagement through digital tools. Future Cities Catapult noted a third emerging trend, though, with citizenship being traded for consumerism. Open data and digital platforms are enabling new business models that blur the lines between citizens and consumers. Airbnb and Uber are two examples of technology-enabled transformative business models that are changing people's daily lives and habits.

Businesses such as IBM, Schneider Electric, CISCO and Siemens have used the concept of a smart city to market their vision for the cities of tomorrow through the 'application

Table 11.1 Industrial definitions of smart cities

Company	Vision	Key vision
IBM	Cities can capitalise on new technologies and insights to transform their systems, operations and service delivery; being smarter can change the way cities work and help deliver on their potential as never before	■ Big data and analytics for deeper insights ■ The 'cloud' for collaboration among disparate agencies, mobile to gather data and address problems directly at the source, social technologies for better engagement with citizens
Schneider Electric	Cities need to become smarter, more efficient, sustainable and liveable – this can be done through collaboration with different entities (municipality, council, etc.) to deliver urban efficiency	■ Smart energy: energy management system to make end-users, renewable energy sources and electric vehicles efficient and permitting smart connection to the grid ■ Smart water: use of management systems to detect water leaks in the network, to optimise the energy used for supplying water, and to provide solutions to face storms and floods ■ Smart building: use of building management systems to monitor energy use ■ Smart mobility: traffic and transit management systems that deliver real-time visibility across the entire transportation network, electric vehicles, and efficient and safe recharging infrastructure via tolling and congestion-charging solutions ■ Smart public services: solutions ranging from street lighting to public safety with a focus on data collection for better management ■ Smart integration: linking different management systems available in the city to increase the efficiency of each one of them and the overall efficiency of the city.
Siemens	Smart cities should find ways to optimise their infrastructure through intelligent infrastructure solutions – such as smart grids, building automation, security solutions and traffic control systems	■ The use of sensors, communications, computational ability and control in some form to enhance the overall functionality of the electric power delivery system

Table 11.1 Continued

Company	Vision	Key vision
Cisco	Smart cities should include an integrated urban ICT that can overlay a city and can support delivery of connected urban services and allow for efficient management of those services on a global scale	■ Leveraging the internet of things (IoT), cities can integrate people, processes, data and things to create safe and vital places to live, work, learn and play

From Bull and Azzenoud (2016).

of complex information systems to integrate the operation of urban infrastructure and services such as buildings, transportation, electrical and water distribution, and public safety' (Paroutis *et al.*, 2014, p. 2). Future Cities Catapult (2017) agreed that global technology companies saw an opportunity to sell digital transformation and new technology into big city systems (water, energy, transport): '"Smart city" caught the imagination as smart phones and digital transformation spread across the world at a phenomenal rate.' However, according to Harrison and Donnelly (2011), this concept is not new, and its origins go back to the smart growth movement in the late 1990s.

Policy-makers have been swift to react to the smart city agenda. Whether at the local, national or European/international level, there is no shortage of guidance, local action and policy directives. Caprotti *et al.* (2016) found examples of nearly a third of UK's towns and cities developing plans for activities that could be labelled 'smart'.

The EU's focus on smart cities is managed through the European Commissions' European Innovation Partnership (EIP). Smart cities have become a major policy initiative of the EU, with the smart city framed as a key vehicle for delivering urban sustainability (Martin *et al.*, 2019). The EU in its Strategic Implementation Plan for 'smart cities and communities' (EC, 2013) defines smart cities as

> systems of people interacting with and using flows of energy, materials, services and financing to catalyse sustainable economic development, resilience, and high quality of life; these flows and interactions become smart through making strategic use of information and communication infrastructure and services in a process of transparent urban planning and management that is responsive to the social and economic needs of society.

In this document it describes areas of focus around sustainable urban mobility, energy-efficient buildings and integrated infrastructures and processes across energy, ICT and transport. Space is given to the need for increased citizen engagement and the benefits that brings. The areas of focus are (*a*) developing a common European framework for cities; (*b*) removing barriers from experimental initiatives that innovate and increase knowledge, and support co-creation; and (*c*) establishing local citizens committees to work with local public authorities, small and medium-sized enterprises (SMEs) and larger industry in order to set the targets for developments. Particular focus is on the establishment of Lighthouse initiatives, which currently fund 14 projects, including 40 Lighthouse cities and 43 'follower' cities. Martin *et al.* (2019) reviewed nine projects according to their digital agendas and sustainability agendas. Table 11.2 summarises these projects, and includes the latest five that have recently been funded.

In the UK the Department for Business, Energy and Industrial Strategy (formerly known as the Department for Business, Innovation and Skills) has defined the process by which cities turn into smart ones (BIS, 2013). It refers to the process as one in which cities become more 'liveable and resilient'. For the UK government, a smart city should enable every citizen to engage with all the services on offer, public as well as private, in a way best suited to his or her needs, and incorporates 'hard infrastructure, social capital including local skills and community institutions, and (digital) technologies to fuel sustainable economic development and provide an attractive environment for all' (BIS, 2013, p. 7). It noted five key features that should underpin a smart city. These are

1. a modern digital infrastructure
2. a recognition that service delivery is improved by being citizen-centric
3. an intelligent physical infrastructure ('smart' systems or the IoT)
4. an openness to learn from others and experiment with new approaches and new business models
5. transparency of outcomes/performance, for example city service dashboards to enable citizens to compare and challenge performance, establishment by establishment and borough by borough.

These are further described in the British Standards Institution (BSI) specification for smart cities 'Smart city framework – guide to establishing strategies for smart cities and communities' (BSI, 2014). In this document the BSI noted that a smart city should be visionary, citizen-centric, digital, and open and collaborative (Figure 11.1). We further note that the notion of digital transparency was raised in the introduction (see Chapter 1).

It is clear, then, that – on paper at least – a purely techno-centric view of smart cities is dissipating. Policy-makers and practitioners are starting to see the citizen is an essential stakeholder, even if there is a blurring over the boundaries between citizens and consumers. It is also unclear what these policy-makers actually refer to when they talk about citizen engagement. There is a world of difference, for example, between simply being informed or consulted and empowered. This is discussed shortly, but first some final reflections on definitions of smart cities.

The shifting definitions of smart cities have been captured well in the academic literature. For example, a comprehensive review by De Jong *et al.* (2015) highlighted 12 different categories of cities in the literature for the period running from 1996 to 2013: 'sustainable

Table 11.2 Overview of the EC Lighthouse initiatives

Lighthouse initiative	Digital agendas		Sustainability goal			
	IoT	Big Data	Digital citizen participation	Greening the urban economy	Reducing greenhouse gas emissions	Promoting social equity
Remourban			×	×	×	
Grow Smarter	×			×	×	
Triangulum	×			×	×	
Sharing Cities	×			×		
Smarter Together	×		×	×	×	
SmartEnCIty				×	×	
Replicate	×			×		
Ruggedised	×			×	×	
mySmartlife			×	×	×	
City Exchange			×		×	
Making a City				×	×	
Stardust		×	×		×	
Match up	×				×	×
IRIS			×	×	×	

Adapted from Martin *et al.* (2019).

Figure 11.1 Vision for a smart city. (Source: after BSI, 2014)

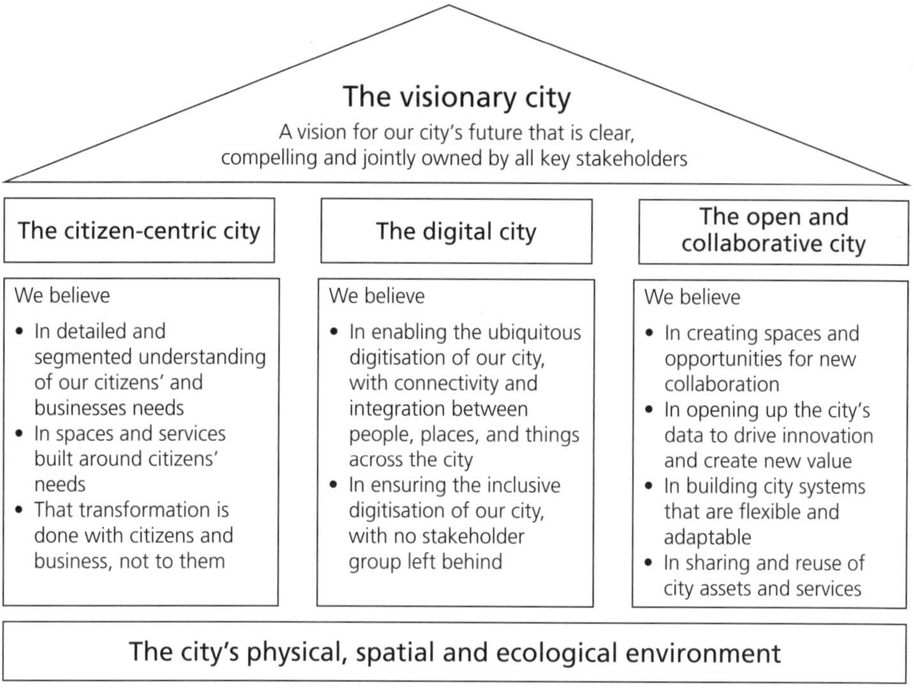

city', 'eco city', 'low-carbon city', 'liveable city', 'green city', 'smart city', 'digital city', 'ubiquitous city', 'intelligent city', 'information city', 'knowledge city' and 'resilient city'. They found 'sustainable city' had the highest number of occurrences, followed by 'smart city'. However, the importance of this study resides in defining the links between these different types depending on their number of occurrences in the selected range of academic literature.

Huber and Mayer (2015) noted that there is no clear definition or conceptual content of smart cities, unlike the low-carbon and eco cities, and that it is still a fuzzy concept; but there exist many interpretations. They conceptualise this through three perspectives

- Instrumental perspective: this consists of using ICT to gather high-quality data from different sources of information in shorter times to help improve the work of institutions, such as municipalities, through the processing of these data in order to produce meaningful information that can help in building the right strategies and making decisions.
- Administrative perspective: the goal of a smart city is to unify the work of institutions through the establishment of a smart policy. In other words, it is fundamental for all structures/departments belonging to the same municipality, as an example, to interact and unify their efforts to develop a vision to the city; a vision that has as a starting point defining the needs of the citizens and as an

end-point meeting these needs. There are parallels here to the ideas of smart governance introduced in Chapters 9 and 10.

- Governance perspective: citizens should have a significant role in defining how their cities should look. This is why it is essential to overcome the traditional top-down governance and transition to a new governance style; a style that enables integration of all stakeholders in the decision-making.

This governance perspective gets to the core issue of how citizens are engaged in decision-making, be it for the design of a new building, infrastructure project or city-level planning such as a new transport policy or carbon management strategy. What does it mean to actually engage the citizens of a particular area or city? Martin *et al.* (2019) agree that 'digitally enabled citizen participation' is a key feature of smart cities (alongside the IoT and big data). This emphasis is also found in policy documents such as the BSI specification for smart cities that have a strong emphasis on the need for citizen engagement, be it by actual face-to-face stakeholder engagement or through the use of digital platforms.

However, other perspectives have raised concerns about some of the underlying assumptions around smart cities. A notable example is that of Martin *et al.* (2018), who feared that these visions of smart cities are ultimately underpinned by capitalist or consumer understandings of cities. Relatedly, consumerist and corporate practices may be subject to accusations of 'greenwash', while internal corporate actions may be constrained by limited notions of legitimacy (see Chapter 10). Undertaking a review of European and North American interpretations, Martin *et al.* (2018, p. 18) concluded that 'smart city initiatives in practice reinforce the focus on delivering unsustainable forms of economic growth and consumerist cultures, while neglecting social equity and environmental protection'. On a more positive note they also noted that these new models of smart city offer greater potential for new models of urban governance, for example innovative partnerships between business, local authorities and citizens groups to facilitate the development of data platforms, citizen engagement and empowerment. Before going on to explore such models, the literature on engagement, and in particular deliberation, is briefly presented.

11.3. The deliberative turn

Much has been made of 'citizen engagement' across the range of definitions discussed above. As noted above, citizen engagement regularly features now in the rhetoric of smart cities. But it is a contested term and can mean different things to different people. The principles of public participation methods have been tried and tested in the siting of controversial facilities such as waste facilities (Bull *et al.*, 2008) and transport planning (Bickerstaff and Walker, 2005). Sovacool (2014) noted three benefits of engaging non-experts: first, democracy is increased, as all citizens have a right to participate and be represented in environmental decision-making; second, non-experts are often more attuned to the ethical issues of a situation; and, third, greater acceptance can often be achieved by involving those affected by the situation. Most relevant to this subject is the strong and emerging links between public engagement and learning, increased environmental citizenship and behaviour change (Bull *et al.*, 2008; Webler *et al.* 1995).

Figure 11.2 Arnstein's ladder of participation

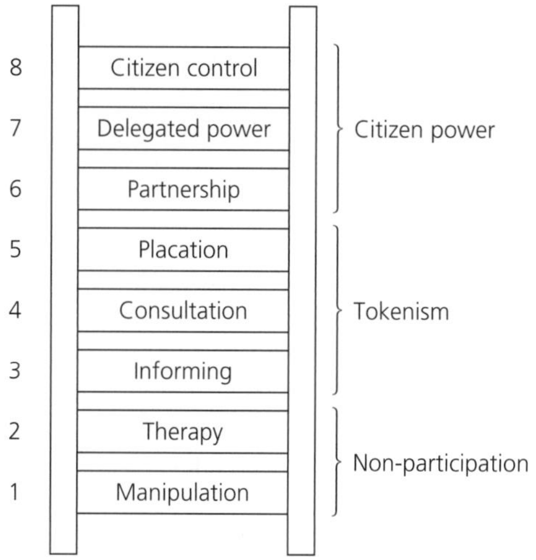

Back in the 1960s, Arnstein's (1969) 'ladder of participation' (Figure 11.2) defined steps to better engagement. At the bottom was information provision as a predominantly one-way form of communication. Moving up the steps, consultation is usually conceived as a relatively passive process asking for people's opinions but not necessarily engaging them in debate. Participation is normally used to refer to processes that allow people to participate in a decision by putting forward their views verbally whereas engagement goes further, suggesting an innovative and interactive two-way process of discussion and dialogue (i.e. deliberation) to ensure that people's views inform a decision, alongside those of the expert and/or decision-maker. This is still one-step removed, however, from Arnstein's top step of her ladder that defines empowerment as people taking control of decisions and their implementation.

The theoretical underpinnings find their roots in Habermas's theory of communicative competence, which was successfully mined in the early 1990s by Webler (1995). Webler (1995) explored how language functions to form key foundational principles for the management of deliberative practices within the school of risk communication. Working from the premise that participation is interaction among individuals through the medium of language (Webler 1995), Habermas (1979) argued that any communication between two individuals would fail without cooperation. An individual's ability to use language to create understanding and consensus is referred to as 'communicative competence'. Habermas (1979) outlined a set of ideal conditions in which communicative competence would be best served, known as his 'ideal speech situation'. Webler (1995) applied these principles of communication to the formulation of a set of criteria and rules that would transform democratic ideals of deliberative democracy into practice.

In short, people can be a valuable source of knowledge and wisdom and, if given the opportunity, capable of handling complex information and resolving complex problems.

The practical realities of implementing meaningful engagement that allows for fair engagement of all through a competent process is a known challenge. Applying this framework to a dynamic and fluid process such as smart cities may be problematic but it is necessary. Notions of citizenship and consumerism are always shifting, and new modes of engagement are emerging. The next section examines some contemporary examples of smart city engagement that are attempting to connect citizens in new and innovative ways.

11.4. Examples of innovative smart city engagement

Public participation normally focuses on the gathering of people to discuss public issues in arenas such as town hall meetings, public hearings, focus groups or community advisory fora (Bull *et al.*, 2008). These traditional methods of public participation require people to physically turn out. But what might citizen engagement look like in smart cities, and are new methods emerging that transcend the need for physical gatherings? As we noted in Table 11.2, six of the 14 Lighthouse city projects have some form of digital citizen participation. This section explores two emerging models of citizen engagement for smart cities. The first, with reference to the EU Lighthouse programme, discusses the example of Nottingham in the UK and the REMOURBAN project, which has prioritised citizen engagement but uses mainly face-to-face approaches. The second, in contrast, discusses three technology-led approaches: one collects ideas from the public, referred to as idea generation, while the other two examples relate to the recommendation of the ideas, and are referred to as idea evaluation (Chiu *et al.*, 2014).

11.4.1 REMOURBAN – 'face-to-face' citizen engagement

REMOURBAN (REgeneration MOdel for accelerating the smart URBAN transformation) is one of 14 Lighthouse smart city demonstrator projects (as noted in Table 11.2), supported by the EU Horizon 2020 investment programme for 5 years (2014–2019). It is a partnership between three Lighthouse cities: Nottingham (UK), Valladolid (Spain) and Eskisehir (Turkey); and two 'follower' cities, Seraing (Belgium) and Miskolc (Hungary). Each partner city aims to develop novel solutions independently, according to its own local needs. These smart city solutions and innovations will then be shared across the five follower cities to develop generic solutions. As a Lighthouse city, Nottingham can offer insights into the role of community engagement as a tool to deliver smart city innovation. The project has three areas of focus – sustainable urban mobility, integrated infrastructure, and sustainable districts and the built environment. Citizen engagement took centre stage for a local demonstration area (Sneinton), where some local residences were retrofitted using the Energiesprong whole-house renovation approach, which included external cladding, a solar roof and a ground-source heat pump. Alongside the retrofitting of homes, investment has gone into supporting the electrification of the city's bus fleet and an innovative car club. The citizen engagement strategy built on the city's past processes, and develops new ideas using the principles outlined in Figure 11.3 and also functions as a pilot to be replicated across the city.

These three levels were broken down into six key practical steps that the project team and the local authority could undertake. These are outlined in Table 11.3.

Figure 11.3 REMOURBAN and citizen engagement (Mazhar *et al.*, 2017)

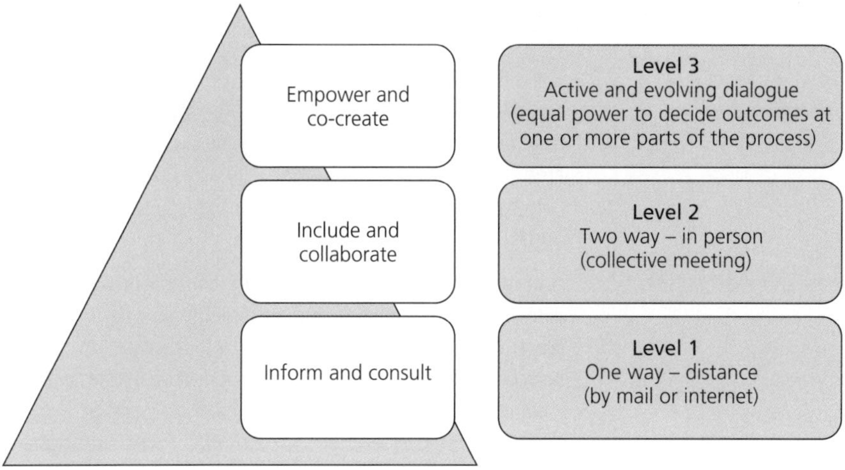

This REMOURBAN methodology of citizen engagement provides cities with a potentially useful model for developing citizen engagement for smart city transformation. As noted in Table 11.3, predominantly face-to-face measures were used. Notable barriers were encountered around a lack of knowledge and understanding by participants, a lack of funding and resources (from the local authority), and challenges around partnership

Table 11.3 REMOURBAN: six steps to citizen engagement in Nottingham

Step	Comment
1. Analysis of the current situation	The REMOURBAN team developed a list of citizen engagement activities for the demonstration area and the whole city via a SWOT analysis. This included ■ direct mail to households and key local influencers such as councillors, MPs, and tenant and community groups ■ local energy events ■ social media ■ press releases to local media.
2. Definition of messages	REMOURBAN defines citizen engagement initiatives as 'processes by which public concerns, needs and values are incorporated into decision-making'. Nottingham developed positive messages for all three levels of citizen engagement for the demonstration area. However, there is a lack of clarity about how these messages are delivered. This suggests that the messages are mainly developed for level 1 and need improvements for more mature levels of engagement.

Table 11.3 Continued

Step	Comment
3. Target audience and expected outreach	The target audience is landlords of privately rented homes, commercial businesses in the demonstrator area, city-wide citizens, community groups and politicians. The demonstration area is a relatively active community and has well-established community groups. This area has a high number of privately rented homes.
4. Tools and mechanisms	A combination of online and offline citizen engagement activities are available, including direct mail, one-to-one visits, community events, news channels, a local newsletter, local noticeboards, community champions, social media, websites, local media, local TV (Notts TV), a local newspaper (*Nottingham Post*) and local radion (BBC Radio Nottingham).
5. Action plan for citizen engagement	Key actions for citizen engagement in REMOURBAN include ■ a stakeholder briefing pack, 'engage the city and Sneinton', targeted information for demonstration houses, and create marketing collateral ■ a citizen engagement implementation plan for energy interventions developed for the demonstration area ■ 465 households segmented into a typology group (e.g. social and private households) to target consultation events, and supporting materials to streamline the process ■ early meetings planned to ensure that people can have their say in the development of the delivery plans – this included a step-by-step 'process map', which detailed the work programme, daily liaison control, regular local events, sign off of the completed work and customer satisfaction.
6. Description of resources	Communications and marketing personnel within the Nottingham City Council's energy services team led on engagement activities. £15 000 was to be spent on the local desk (a marketing officer in the energy services team) placement and marketing collateral in the project. Beyond the project, though, there was a lack of funding to effectively implement projects.

Adapted from Mazhar *et al.* (2017).

working (Mazhar *et al.*, 2017). Given the challenges of these resource-intensive and large-scale engagement activities, it is interesting to consider other smaller-scale examples that offer complementary insights into how citizens can engage with change at a local and city scale.

11.4.2 Examples of digital citizen engagement

This section provides an overview of three smaller-scale examples of digitally enabled engagement that begin to blur the boundaries of citizen- and consumer-led engagement: crowdsourcing, online feedback, and voting and gaming. Firstly, crowdsourcing is an online, distributed problem-solving model that is used in smart cities to collect information on a wide range of topics (Brabham, 2008). The work, in this case providing ideas of how to improve the city, is done by a large group of people, in this case the citizens, and the topics can extend to any component of the city. For example, the Spanish city of Zaragoza has developed online tools to receive feedback on the local public infrastructure (Aguilera *et al.*, 2017). The city provides a list of the reports and faults that describe the current situation and ask the local citizens for new complaints and suggestions – FixMyStreet is often mentioned in this connection. The information that is crowdsourced from citizens can relate to any element of the public infrastructure and in this case the location of the comment is automatically attached to the feedback form through the geo-location capability of the device that the citizen uses.

The second digital smart city approach is to request feedback or voting on a narrow range of topics that are of particular interest. In this case the problem is related to a specific component of the city, and the feedback is once again given by the crowd. For example, the Spanish city of Castellón has developed an online feedback tool that allows users to give feedback regarding the city's bike-sharing facilities (Aguilera *et al.*, 2017). This was considered to be a suitable topic for feedback, as the city had already created an app that reported the real-time availability of the bicycles at each station. The app allows the users to report issues relating to the sharing services and the state of the bicycles. Similarly, Maptionnaire can be used to collect feedback on a specific range of topics. It allows organisations to quickly create their own online questionnaires, and the questionnaires can then easily be linked to the areas of interest by maps based on a geographic information system (GIS). This enables the feedback to relate to specific area of the city and to report on how it can be improved. The tool also has the capability to analyse the feedback and to store all the collected citizen feedback in a common database.

Citizens can also engage with their city government on a specific set of topics via voting. This allows the public to read about a range of options and to recommend the option that they consider to be the best fit. An advantage of this approach is the speed with which citizens can participate, as it does not require qualitative feedback. For example, the city of San Francisco uses the MindMixer platform and encourages participation through civic-based rewards. In one scenario it was used to select a new logo for the San Francisco Municipal Transportation Authority (Seltzer and Mahmoudi, 2013).

Finally, in some cases serious games have been used to receive feedback from the public (Poplin, 2012). One urban planning example from Germany concerned the renovation of a number of university buildings. There were four possible options for the project, which included renovation of the existing buildings, demolition and new construction of the existing buildings, partial relocation to the new site or complete relocation to the new site. The goal of the game was to encourage the public to study the alternative options in detail rather than just speed reading through the proposals. In order to be successful at the game the players needed to understand the advantages and disadvantages of the various options and to make trade-offs when selecting their preferred option. The game communicated the trade-offs and enabled the players to find the urban planning solution that was most acceptable to them.

The use of such online tools clearly reduces one of the key barriers to traditional public participation exercises; that is, actually persuading people to attend physically. They reduce the cost that citizens incur when giving feedback, and the main cost that is reduced is time. The citizen is not required to report the feedback in person at an official location or at a community meeting, and the expected result is that the public bodies will receive a greater amount of feedback. There are two reasons why the amount of feedback is increased. Firstly, they may increase the total number of citizens that participate in public dialogue and, secondly, they may increase the frequency at which they participate.

11.5. Discussion

In the context of public participation, the role of the citizen can be divided into two categories – the political actor and the consumer (Anttiroiko, 2016). The actions of the political actor match the agenda of traditional public participation, as they include policy-making, planning or governance processes, whereas the consumer typically participates in a facilitated user-driven innovation process. This definition of consumer echoes one of the aims of the smart city framework that has been developed by the BSI, which is to enable citizens to not just be users of services but to have a specific and active role in the transition (BSI, 2014).

Many of these new models of smart city engagement shift the whole emphasis of engagement from an active choice that citizens have to make (in the case of REMOURBAN, for example: 'Do I attend this event?') to an integrated one in which, by their very interaction with the services offered by the smart city, citizens are providing feedback. So, as the smart city consumers move through a smart city, they interact with the various services by using public transport, hailing a taxi, visiting a gym or buying lunch, and they leave a digital footprint. In doing so they are acting as mobile sensors, and the data generated can be used to take earlier and better decisions and to provide better services (Aguilera *et al.*, 2017). This is important in the current marketplace, as consumers are increasingly demanding complex, sustainable and integrated solutions rather than standardised and homogeneous products and services (Parente *et al.*, 2018). Unlike the political actors, the consumers are not actively participating in public dialogue. Their motivation is to communicate their preferences and needs in the hope that their digital footprint will shape the products and services that are on offer in the smart city. This feedback is effortless and continuous. It can be regarded as passive participation.

The access economy has the potential to meet the needs of these new customer expectations, and the digital footprints of the smart city consumers play an intrinsic role in the development of the service offerings. The access economy, also commonly called the sharing economy, includes businesses such as car sharing (Zipcar) and space sharing (WeWork), where a firm's assets are temporarily rented to consumers (Parente *et al.*,

2018). The access economy also delivers some of the sustainability targets of the smart city, as the re-use can ease the pressure on natural resources. It is also thought that the shift from individual ownership to collaborative consumption can reduce hedonistic consumerism and provide a sustainability framework based on community sharing (Ganapati and Reddick, 2018).

For example, WeWork is a fascinating case of an access economy business where the service offering is continuously reconfigured based on the digital footprint of the consumers. The company was founded in 2010, and it offers shared workspace and related services to a quarter of a million members across 75 different cities (*The Architects Newspaper*, 2018). Buildings have typically been designed as bespoke projects where the assortment of spaces in a building is selected to meet the needs of the users. Once the interior walls are constructed, they are not expected to be demolished or moved for years to come. As with a family home, you cannot reconfigure the floorplan frivolously. If, for example, a particular office building is designed to have a boardroom for 20 people and ten small meeting rooms, then that will be the space offering available to the users of the building in the short to medium term. In contrast, WeWork has challenged this way of thinking, and it is constantly tweaking and updating its space offering based on user feedback. When the users want a desk or a meeting room to work in, then they reserve the desired space via an online reservation tool, and this creates a detailed digital footprint. If the building has several alternative open office areas with themes such as a quiet library zone, a noisy coffee house zone and a jazz music zone, then the users can choose the space that suits their working styles on that particular day. The data gathered by sensors and the reservation system can then show which spaces the users need more of and which spaces they need less of (*The Architects Newspaper*, 2018). If there are ten phone boxes for making private phone calls and the data show that the peak occupancy at any one time is five phone boxes, then the chances are that five of them will be removed and replaced with a workspace that is more in demand.

A recent study on the design of workspaces has argued that achieving both lower costs and higher productivity requires taking a data-driven and holistic view of the workplace (Lees, 2018). Apart from the commercial benefits of cost and efficiency, there is also a case to be made that a continuous reconfiguration of the offering provides a citizen-centric set of services. This is because the remaining offering has been optimised based on the citizens' needs, and thus it is giving the citizens exactly what they want. The same study also argued that the use of data and evidence to drive decision-making should not be confused with manipulation by management. Instead, it is the opposite of this, as better data enable the decentralisation of managerial control (Lees, 2018). This begs the question of whether passive participation by smart city consumers the first steps towards citizen control as defined by Arnstein (1969) or is there still a long journey ahead. The relationship between the smart city and the citizen is a theme that recurs throughout this book, having first been introduced in Chapter 1.

11.6. Conclusion

These two contrasting examples of engaging citizens present an interesting, shifting and important picture of how citizens are currently involved, or not, in smart city developments. Firstly, it is important to be clear what is being talked about with regards to smart cities, what is promised with regards to these developments and who is defining the terms of engagement for smart cities. It is neither desirable nor democratic for the large ICT players to shape our cities without the clear engagement and involvement of the people who live in them, especially given legitimate concerns around data privacy and security, as these new data-sharing platforms, from smart meters to sharing platforms, harvest personal information. Engagement models are changing, though. The case study of Nottingham shows the challenges and scale in organising face-to-face citizen engagement. Though ultimately it may be viewed as the 'gold standard' – recent awards for the retrofitting that has been undertaken are testament to that – our other examples of more-disruptive tools show that there is not a one-size-fits-all solution. These newer and distributed examples of how citizens engage by their daily activities has shown that citizen engagement may in the future be embedded in the very essence of the new technologies and services and no longer viewed as a separate activity. Smart cities, and the citizens inhabiting them, are blurring the lines between political actors and consumers, and engagement must be more than providing information and feedback. Genuine participation that accesses the knowledge and skills of all the actors and stakeholders is vital to provide greater legitimacy and acceptance of new low-carbon solutions.

REFERENCES

Aguilera U, Peña O, Belmonte O and López-de-Ipiña D (2017) Citizen-centric data services for smarter cities. *Future Generation Computer Systems* **76**: 234–247.

Anttiroiko A (2016) City-as-a-platform: the rise of participatory innovation platforms in Finnish cities. *Sustainability* **8**: 922–953.

Arnstein SR (1969) A ladder of citizen participation. *Journal of the American Planning Association* **35(4)**: 216–224.

Bickerstaff K and Walker G (2005) Shared visions, unholy alliances: power, governance and deliberative processes in local transport planning. *Urban Studies* **42(12)**: 2123–2144.

BIS (Department for Business, Innovation and Skills) (2013) *Smart Cities: Background Paper*. BIS, London, UK. https://www.gov.uk/government/uploads/system/uploads/attachment_data/file/246019/bis-13-1209-smart-cities-background-paper-digital.pdf (accessed 27/04/2015)

Brabham DC (2008) Crowdsourcing as a model for problem solving: an introduction and cases, convergence. *International Journal of Research into New Media Technologies* **14(1)**: 75–90.

BSI (British Standards Institute) (2014) PAS 181:2014. Smart city framework. Guide to establishing strategies for smart cities and communities. BSI, London, UK.

Bull R and Azennoud M (2016) Smart participation- social learning: a model of participation. *Proceedings of the Institution of Civil Engineers – Energy*, 10.1680/jener.15.00030.

Bull R, Petts J and Evans J (2008) Social Learning from Public Engagement: Dreaming the impossible? *Journal of Environmental Management and Planning* **51(5)**: 703–718.

Caprotti F, Cowley R and Flynn A (2016) *Smart-Eco Cities in the UK: Trends and City Profiles 2016*. University of Exeter, Exeter, UK.

Chiu C, Liang T and Turban E (2014) What can crowdsourcing do for decision support? *Decision Support Systems* **65**: 40–49.

Cowley R and Caprotti F (2018) Smart city as anti-planning in the UK. *Environment and Planning D: Society and Space*, 10.1177/0263775818787506.

De Jong M, Joss S, Schraven D, Zhan C and Weijnen M (2015) Sustainable smart resilient-low carbon eco knowledge cities: making sense of a multitude of concepts promoting sustainable urbanization. *Journal of Cleaner Production* **109**: 25–38.

EC (European Commission) (2013) European innovation partnership on smart cities and communities: strategic implementation plan. https://ec.europa.eu/info/eu-regional-and-urban-development/topics/cities-and-urban-development/city-initiatives/smart-cities_en#european-innovation-partnership-on-smart-cities-and-communities (accessed 16/03/2015).

Future Cities Catapult (2017) *Smart City Strategies: A Global Review 2017*. Future Cities Catapult, London, UK. https://futurecities.catapult.org.uk/project/global-review-smart-city-strategies/ (accessed 20/06/2017).

Ganapati S and Reddick CG (2018) Prospects and challenges of sharing economy for the public sector. *Government Information Quarterly* **35(1)**: 77–87.

Habermas J (1979) *Communication and the Evolution of Society*. Beacon Press, Boston, MA, USA.

Harrison C and Donnelly IA (2011) A theory of smart. *55th Annual Meeting of the ISSS, Hull, UK*, pp. 1–15. http://journals.isss.org/index.php/proceedings55th/article/viewFile/1703/572 (accessed: 01/03/2015).

Huber A and Mayer I (2015) Is this a smart city? Narratives of city smartness and their critical assessment. *Proceedings of the 2015 ECEE Summer Study Conference, Buffalo, NY, USA*, pp. 817–823.

Janda K (2011) Building's don't use energy: people do. *Architectural Science Review* **54**: 15–22.

Lees C (2018) Delivering and managing high productivity, low cost workplaces: a data driven perspective. *Corporate Real Estate Journal* **7**: 3.

Martin C, Evans J and Karvonen A (2018) Smart and sustainable? Five tensions in the visions and practice of the smart-sustainable city. *Technological Forecasting and Social Change* **133**: 269–278.

Martin C, Evans J, Karvonen A *et al.* (2019) Smart-sustainability: a new urban fix? *Sustainable Cities and Society* **45**: 640–648.

Mazhar M, Kaveh B, Sarshar M and Bull R (2017) Community Engagement as a tool to help deliver smart city innovation: a case study of Nottingham, United Kingdom. *Proceedings of the 2017 ECEEE Summer Study Conference, Île de Giens, France*, pp. 807–820.

Parente RC, Geleilate JG and Rong K (2018) The sharing economy globalization phenomenon: a research agenda. *Journal of International Management* **24(1)**: 52–64.

Paroutis S, Bennett M and Heracleous L (2014) A strategic view on smart city technology: the case if IBM Smarter Cities during a recession. *Technological Forecasting and Strategic Change* **89**: 262–272.

Poplin A (2012) Playful public participation in urban planning: a case study for online serious games. *Computers, Environment and Urban Systems* **36(3)**: 195–206.

Seltzer E and Mahmoudi D (2013) Citizen participation, open innovation, and crowdsourcing: challenges and opportunities for planning. *Journal of Planning Literature* **28(1)**: 3–18.

Sovacool B (2014) What are we doing here? Analyzing fifteen years of energy scholarship and proposing a social science research agenda. *Energy Research and Social Science* **1**: 1–29.

The Architects Newspaper (2018) WeWork is using user data to chart their meteoric expansion. https://archpaper.com/2018/07/wework-data-meteoric-expansion/ (accessed 24/10/2018).

UN (United Nations) (2018) *The Sustainable Development Goals Report 2018*. UN, Geneva, Switzerland. https://unstats.un.org/sdgs/report/2017/goal-11/ (accessed 20/06/2019).

UNEP (UN Environment Programme) (2015) *Cities and Buildings*. UNEP-DTIE Sustainable Consumption and Production Branch, Paris, France. https://issuu.com/rodrigovelasquezangel/docs/cities_and_buildings-unep_dtie_init/4 (accessed 01/10/2015)

Webler T (1995) 'Right' discourse in citizen participation: an evaluative yardstick. In Fairness and Competence in Citizen Participation: Evaluating Models for Environmental Discourse (Renn O, Webler T and Wiedemann P (eds)). Kluwer, London, UK, pp. 35–86.

Webler T, Kastenholz H and Renn O (1995) Public participation in impact assessment: a social learning perspective. *Environmental Impact Assessment Review* **15**: 443–463.

Conclusion

ice Publishing

Nuttall, William J, Gibson, David V, Trzmielak, Dariusz and Ibarra-Yunez, Alejandro
ISBN 978-0-7277-6425-6
https://doi.org/10.1680/emsc.64256.231

Chapter 12
Conclusions

Alejandro Ibarra-Yunez
Tecnológico de Monterrey, Mexico

William Nuttall
The Open University, Milton Keynes, UK

David Gibson
The University of Texas at Austin, USA

Dariusz Trzmielak
University of Łódź, Poland

12.1. Smart cities in a stressed world

Smart cities have become well-planned projects involving government civil servants, urbanists, information and communications technology (ICT) experts, city dwellers, and companies in charge of a myriad of goods, services and applications. At its best the smart city vision is oriented to the improvement of life, health, mobility and the environment as well as the well-known R-R-R actions: recycling–reuse–reduce. With that said, the smart city has not been the consequence of a single initiating idea, but rather the concept has first emerged by serendipity in many places around the world. Arguably the first elements of the concept came from the potential of ICTs to improve communication in government services, but also an impetus came from a more general and dramatic growth in computer use. We note particularly the use of computing for life improvement accompanied by a reduction of the cost of both data gathering and management.

The global population has grown significantly since the beginning of the 20th century. In the first half of that century the annual rate of population growth accelerated to a peak of 2.1% around 1970, before slowly dropping back (Roser and Ortez-Ospina, 2019). Despite a slowing rate of growth, the global population is still increasing. This continued growth has become a subject of concern to numerous organisations. The distribution of populations is also shifting, the United Nations Population Division predicts that around 68% of the world population will live in urban areas by 2050, while the equivalent figure was 50% in 2006, and just 26% in 1960. Services in cities have become strained, and this is a source of serious concern for governments and regulatory agencies, such as those responsible for urban planning, health planning and environmental quality management. These issues are also of concern to those concerned with economic development. Meanwhile, services for the increased population are also under strain, and that pressure also extends to companies, financial institutions, ICT providers and all stakeholders simply sharing their space with others in the new smart cities.

12.2. The developing world

Throughout this book we have tended to concentrate on activities in some of the richest and most economically developed countries in the world. At the outset of this project that was not our intention. The focus emerged from the global distribution of innovation, or at least of visible innovation. We are sure that there are smart-city-related innovations emerging in all parts of the world. In particular, we are drawn to the nascent concept of the 'smart village'. Such developments have greater potential for leadership to be seen in less-developed economic contexts. One can already observe interesting developments concerning off-grid connectivity, especially underpinning improvements in education, e-healthcare and communication: see, for example, the observations of the European Network for Rural Development (ENRD, 2019). We welcome such innovations, but it is not yet clear how they might influence the direction of travel visible in the smart cities considered earlier in this book. Pointing to one possible example of developing country leadership, one can consider the deployment of mobile-phone-based financial services in countries lacking fixed-line communication systems (BBVA, 2015). While not without problems, the leap-frogging of technological barriers has shown the way to more developed economies. We are also conscious, however, that, for example – in the context of renewable energy innovation – Chapter 7 of this book reports that

> The level of economic development also helps explain the time when wind and solar PV production began, with most Latin American economies in the sample, plus China and India, beginning much later than their developed counterparts.

In Chapter 8 we hear about the role of smart technologies supported by agent-based modelling (ABM) in our energy future. As the authors of the chapter point out, such opportunities are not just restricted to developed countries, such as the UK, but might also be deployed in the developing world. The authors observe that the ABM approach can be deployed internationally with relative ease. They observe

> In this type of ABM, the physics of a dwelling, the particular demands that are needed in the dwelling and the algorithms used in the smart control system can be coded in software and a wide variety of conditions tested in a 'virtual lab' without the need for costly and time-consuming empirical studies in each new community.

12.3. Origins of the smart city

Since the times of Cold War satellite surveillance for defence, ICT has been used as a source of critical information. The processes of ICT have involved data gathering often with significant granularity. Today in civilian life we see the benefits of, and confront the challenges arising from, 'big data'. Today's modern digital communications merged with information technologies can be traced back to work at one university, the City University of New York (CUNY). In 1981, CUNY created the BITNET with assistance from Yale University (Indiana University, 2018). It soon spread out unstoppably through the mid-1980s. The new technology increased the number of traditional telephone conversations, but it also made possible wholly new endeavours, such as telemedicine. In that decade, information and communication capacities increased to develop projects for security alarms in buildings directly connected to the police or private security services. Then buildings themselves became digital, with intelligent elevators and escalators, responsive illumination and new smart heating and cooling. Even windows became smart with variable transparency and thermal attributes. Generally, there has been increasing compatibility and complementarity between the various technologies and applications yielding the creation of the newly intelligent or 'smart' buildings. While BITNET itself faded away in the early 1990s, primarily owing to the rise of TCP/IPTCP/IP-based networking (the internet), it had already established a paradigm by which research universities and other seats of knowledge and learning aspired to communicate to advance their purposes, both scholarly and social. Of course, these universities and institutes tend to be in major cities. They operate and work there, but also their network-aware employees live there too. We posit that BITNET had an early role in the slow emergence of the concept of the smart city.

Increasingly, technologies have been furthered for a very wide range of government services and responsibilities, including intelligent or sensor-driven street lights, surveillance security cameras, transit management, waste management, electricity supply, and even water use and management. Often these challenges are accompanied and propelled by the growth in the size of the city with increasing population agglomerations. It is not just static services that are changing – we see the appearance of smart vehicles with sensors to improve drivers' habits or even self-parking. We now see new public transportation systems that minimise carbon dioxide production. Much of the progress involves electric vehicles benefiting from new battery technologies. Such mobility trends and ideas relate to matters discussed in Chapters 2, 5, 6 and elsewhere in this book. These various innovations are increasingly interacting and synergising, becoming what we might regard as 'meshed technologies'.

12.4. Smart city dynamics

Throughout this book various authors have grappled with the question, 'What makes a city smart?' In Chapter 1 we introduced the work of the Cities in Motion team at IESE Business School based in Navarra, Spain, in mapping the dynamics of change around the smart city concept. The IESE team has considered the actors and areas required to call a city 'smart'. As summarised in Chapter 1, it is suggested that they are

- human capital
- social cohesion
- economy
- governance
- environment
- mobility and transportation
- urban planning
- international outreach
- technology.

That is, these eight dimensions are critical for a smart city. Furthermore, IESE has established metrics for qualifying cities. From them, the main smart cities can be found in Europe (and indeed in this book we have tended to focus on European examples), but there are also important examples in the Americas, and Asia/Australasia. The CITI I/O web resource has pointed to, and summarised, the work of IESE described above (CITI I/O, 2018). In passing, we can also add that CITI I/O is itself an interesting forum, as it stresses the nexus of cities, technologies and people. As such, it is interested in a similar convergence of issues to that which has inspired this book.

The concept of 'smart' has moved in multiple directions. There has been the emergence of meshed technologies applied to controllable settings, as can be seen in modern buildings. There has also been an emergence of extended technologies improving the human experience in villages and more urban environments. These extended technologies can involve commerce, government services and residential information. They can enhance the daily experiences of citizens by improving many aspects of the quality of life in public areas, for example in parks or at transport hubs (e.g. bus stations). In addition, extended technologies can improve the delivery of information to citizens beyond merely geographic considerations. These technologies can improve the provision of financial services, among many other information-based aspects of modern life.

12.5. People

Fundamental to this book has been the premise that the smart city is not a technological innovation per se, and certainly not that it is some kind of new-urban-machine. The smart city is a social community. It is defined by its people more than it is defined by its technologies. Central to the idea of a smart city is that the concept and creativity are always evolving new applications change social interaction. That is, the smart city, as a concept, is a social thing and it cannot be thought of as static. Its dynamic nature is a key aspect of its definition.

Looking to the future, we expect that citizen engagement and the relationship between the individual and the collective (privacy, etc.) will be key to shaping the future direction of the smart city. In particular, and as was rightly posited in Chapter 5: 'Public policy needs to engage and develop more of a partnership approach with the new and increasingly powerful digital actors.'

For some, urbanisation has become a critical matter and a source of welfare degradation, while others see improvements in cities becoming modern, clean and sustainable, with better life quality and with opportunities for innovation and creativity. As has been encountered at several points in this book, even the concept of the smart city has multiple definitions and differing views as to the full extent of the concept. Consequently, this book has referred to definitions and analysis of the smart city that have been varied and evolving. Additionally, new issues are continuously emerging, and the new aspects of energy, mobility and smart cities are arising that can offer the reader food for thought. The authors of the chapters of this book have sought to advance understanding via a framework of knowledge and creative thinking that is pertinent and connected to the various dynamic realities. The topic of the book arose from a conference series: the International Conferences on Technology Policy and Innovation. The inspiration for the book came particularly from the 15th such conference held at The Open University in Milton Keynes, UK, in June 2015. Ideas were further advanced in the context of the 16th conference held in Taipei, Taiwan, in 2017.

12.6. To close

The topic of the book arose from a conference series: the International Conferences on Technology Policy and Innovation. The inspiration for the book came particularly from the 15th such conference, held at the Open University in Milton Keynes, UK, in June 2015. Ideas about smart cities were further advanced in the context of the 16th conference, held in Taipei, Taiwan, in 2017.

As has been encountered at several points in this book, the smart city has multiple definitions and there are different views as to the full extent of the concept. Consequently, this book has referred to definitions and analyses of the smart city that have been varied and evolving. Additionally, these issues are continuously emerging, with new aspects of energy, mobility, governance and community coming into play. The authors of the chapters of this book have sought to advance understanding via a framework of knowledge and creative thinking that aims to be both pertinent, and connected to, the dynamic realities.

This book has been presented organised into four distinctive parts stressing, in turn, *communities*, noting locational opportunities for environmental entrepreneurial efforts; *mobility*, with emphasis on autonomous smart vehicles and urban mobility, innovations in logistics and transport, with successful cases mainly in Europe; *energy*, with comparative analyses concerning the state of government institutions incentivising investments in renewables and their use in urban communities, but also focusing on stakeholders and community agents to promote urban benefits of human cooperation; and *governance*, with emphasis of the critical role of the individual and community systems that together form the smart cities' interconnected dimensions. As editors, we hope and trust that by presenting the evolving concept of the smart city in this way, we have managed to reveal key themes and explore the issues underlying successful smart city developments.

REFERENCES

BBVA (2015) Why are mobile payments so successful in developing countries? https://www.bbva.com/en/mobile-payments-successful-developing-countries/ (accessed 12/05/2019).

CITI I/O (2018) The top 50 smart cities in the world 2018. https://citi.io/2018/07/27/the-top-50-smart-cities-in-the-world-2018/ (accessed 12/05/2019).

ENRD (European Network for Rural Development) (2019) Smart villages. https://enrd.ec.europa.eu/enrd-thematic-work/smart-and-competitive-rural-areas/smart-villages_en (accessed 12/05/2019).

Indiana University (2018) What was BITNET, and what happened to it? https://kb.iu.edu/d/aaso (accessed 12/05/2019).

Roser M and Ortez-Ospina E (2019) World population growth. *Our World in Data.* https://ourworldindata.org/world-population-growth (accessed 12/05/2019).

ice Publishing

Nuttall, William J, Gibson, David V, Trzmielak, Dariusz and Ibarra-Yunez, Alejandro
ISBN 978-0-7277-6425-6
https://doi.org/10.1680/emsc.64256.237

Index